green space, green time

Connie Barlow

green space, green time

the way of science

COPERNICUS
AN IMPRINT OF SPRINGER-VERLAG

Reprinted in 2003 by Connie Barlow,
Connie@TheGreatStory.org
www.TheGreatStory.org

Published in the United States by Copernicus,
an imprint of Springer-Verlag New York, Inc.

Copernicus
Springer-Verlag New York, Inc.
175 Fifth Avenue
New York, NY 10010

Library of Congress Cataloging-in-Publication Data
Barlow, Connie C.
Green space, green time : the way of science / Connie Barlow.
p. cm.
Includes index.
ISBN 0-387-94794-9 (hardcover)
1. Nature—Religious aspects. 2. Human ecology—Religious aspects. 3. Religion and science. I. Title.
BL435.B37 1997
291.1'78362—dc21 97-15704
CIP

Manufactured in the United States of America.
Printed on acid-free paper.

9 8 7 6 5 4 3 2 1

ISBN 0-387-94794-9 SPIN 10539263

In memory of giant ground sloths
In celebration of cottonwood trees

Contents

Acknowledgments *ix*

Portrait Gallery *xiii*

1 The Way of Science 1

A Surprise from Sociobiology 3

Varieties of Ecoreligious Experience 10

Science and Meaning 16

2 Science and the Coming of a New Story 23

The Epic of Evolution 30

Science into Story 44

A Conversation with Catalysts 57

Celebrating the Epic 79

Conversations with Edward O. Wilson, Ursula Goodenough, Brian Swimme, Loyal Rue, Mary Evelyn Tucker, John Grim

3 Biology and the Celebration of Diversity 85

Re-Storying Biodiversity 87

Islands Everywhere 97

A Conversation with an Earth Ecstatic 104

Life Loves Life 108

Conversations with Diane Ackerman, Edward O. Wilson, Stephan Harding, Lynn Margulis, Paul Mankiewicz, Paul S. Martin, Lee Klinger

4 Ecology and the Birth of Bioregionalism 121

Keystones, Aliens, and Ghosts 126

Polarities 140

A Conversation with Hands-Off Stewards 146

Creating Science-Based Rituals 156

Conversations with Paul S. Martin, Paul and Julie Mankiewicz

5 Geophysiology and the Revival of Gaia 159

Powers of Persistence 166

A Developing Biosphere 182

A Conversation with the F_1 and F_2 Generations 201

Beyond the Science 214

Conversations with Jim and Sandy Lovelock, David Schwartzman, Tim Lenton, Tyler Volk, Lee Klinger, Andy Watson, Lee Kump, Stephan Harding

6 Meaning-Making 223

A Federation of Meaning 228

Value-Making 235

Meaning-Makers 261

Living Dangerously 272

Conversations with Dick Holland, John Maynard Smith, Loyal Rue, Ursula Goodenough, Tyler Volk, Julian Huxley

Endnotes 297

Index 319

Acknowledgments

In my case it takes a community to raise a book. I am deeply indebted to all who joined me in conversation and who graciously authorized publication of not only well-formulated ideas but also the fresh insights, passionate outbursts, and spontaneous humor that conversation magically evokes. The names of these interlocutors you will find in the table of contents; their faces, in the portrait gallery. Because of them, assembling this book has been a joy; editing and re-editing chapters would have become solipsistically tedious were it not for the diverse voices that I encountered each time with growing delight.

I am grateful, in addition, to those who offered advice and corrected errors in more than their own parts of the conversations: Dick Holland, Tim Lenton, Lynn Margulis, Paul Martin, and Tyler Volk. I thank Brian Swimme for uncommon encouragement, and Loyal Rue and Ursula Goodenough for e-mail companionship that keeps me conscious of the larger enterprise that engages us all. I thank, as well, my editors—Tyler Volk at home and Bill Frucht at Copernicus Books. Criticism that sent me back to the drawing boards more times than I care to remember is now cherished in hindsight. Bill not only saw the potential in this book at a very early stage; he sharpened its scientific focus while remaining open to innovations that appeared along the way. As a wordsmith, he is a master of deletion.

Others at Springer-Verlag whose assistance has been essential are Victoria Evarretta and Teresa Shields. Connie Day expertly proved the dictum "Even editors need an editor"; she is a master of insertion, substitution, and transposition. Thanks, too, to all who have toiled namelessly behind the scenes. I am especially grateful to those who rendered a detail of Alexis Rockman's gorgeous painting into a book jacket that captures the mood of the text.

Outside reviewers whose comments and pointed queries helped me embed the final chapter within the expansive new field of environmental philosophy include John Davis, Holmes Rolston, and Baird Callicott. I thank John Davis also for his careful editing of the first chapter, and for advance publication of extracts in the Fall 1996 and Spring 1997 issues of the journal he edits, *Wild Earth*. Stephen J. Cross and Crispin Tickell gave me pointers on how to keep the late Julian Huxley in character in my imagined dialogues. Arlon Tussing pressed me to face up to the need for credos. I thank Betsy Koenigsberg, Carl Zimmer, Diane Ackerman, Michael Rampino, Caspar Henderson, and Ann Marek for helping me secure miscellaneous facts, and Ben Volk for his computer wizardry at photograph manipulation. I am grateful, as well, for those who conceived and organized two conferences that proved crucial for generating many of the conversations that appear in this book: the 1996 Gaia conference at Oxford University (organized by Jim and Sandy Lovelock; hosted by Crispin Tickell) and the 1996 Epic of Evolution conference (organized by Loyal Rue and Ursula Goodenough for the Institute on Religion in an Age of Science).

I thank John Surette and staff at the Spiritearth retreat in Saugerties, New York, for hospitality and use of their excellent library when I commenced this project. Later, the interlibrary loan department of Western New Mexico University made it possible for me to access written treasures while enjoying a rural lifestyle. I acknowledge, too, the friendships of this little community along the Gila River, where courtesy and warmth are extended. I acknowledge the enormous gifts of the wildlands of the Upper Gila Watershed, which provided instruction and immersion during the writing of this book. Finally, I acknowledge the trees whose once-living fibers now support this sharing of ideas.

Portrait Gallery

Edward O. Wilson with "Dacie," the metallic sculpture of one of his favorite ant species, Daceton armigerum *of South America. (Photo by Jon Chase)*

Loyal and Marilyn Rue at morning chapel service on Star Island during the 1996 Epic of Evolution conference.

Brian Swimme (gesturing) with Billy Grassie at the Epic of Evolution conference, Star Island.

John Grim (far left) and Mary Evelyn Tucker (far right) at the Epic of Evolution conference.

Brian Swimme.

John Maynard Smith.

Dick Holland with granddaughter, Esther Holland.

Ursula Goodenough with daughter, Jessica Goodenough Heuser.

Diane Ackerman. (Photo by Jill Krementz)

Lynn Margulis.

Paul Martin with mammoth skeleton at the Smithsonian, 1987.
(Photo by Chip Clark)

Paul and Julie Mankiewicz with children Phoebe and Tighe, constructing a miniature wetland in their backyard in the Bronx. (Photo by Richard DeWitt)

Stephan Harding (right) talking with Jim Lovelock during the 1996 Gaia conference at Oxford University.

Tim Lenton and Sandy Lovelock at the 1996 Gaia conference.

James Lovelock. (Photo by Sandy Lovelock)

Lee Klinger and David Schwartzman at the 1996 Gaia conference.

Lee Klinger in the phytotron, a controlled-environment greenhouse, at the National Center for Atmospheric Research, in Colorado. Klinger directs research on Gaia and complexity theory research in this facility.

David Schwartzman with his second favorite organism.
(Photo by Ford Cochran)

Lee Kump holding a core tube while conducting research in a Pennsylvania marsh.

Andy Watson in the Galapagos Islands in 1994, after the first ocean experiment to test whether addition of iron to the "ocean desert" west of the islands would increase marine productivity. The tee shirt, from a previous expedition, shows the path of the global deep-water circulation. Cactus and iguana on the left. (Photo by Kim Van Scoy)

Tyler Volk breathing with the biosphere. (Photo by Connie Barlow)

Connie Barlow with her current favorite organism —
a cottonwood tree along the Gila River in New Mexico.
Note the beaver-cropped willows (and wood chips) surrounding the tree;
protective chickenwire is faintly visible encircling its trunk.
(Photo by Tyler Volk)

1

The Way of Science

Schools, churches, and the media are all beginning to sound the same message. We, and especially the children, hear more and more that it is our duty to live lightly on the land, to refashion our lives in at least modest ways to make amends for the environmental woes we are causing. It can all be quite depressing. We begin to think of our own species as uniquely malign. Only through self-abnegation and saintly acts can we halt the destruction. The best we can hope for is to do less harm. And we do less harm by way of duty.

Duty is that which we do because we should—not because we would. Dutiful acts draw praise or relieve us from the criticism of others. Better, they give us peace of mind, even satisfaction. In contrast, an action that is as natural as drawing breath is not a duty. That which appears to others as a dutiful act but which is undertaken without any thought of duty, perhaps even joyfully, is, rather, a beautiful act. This is Immanuel Kant's terminology. A beautiful act arises from our deepest inclinations. We simply could not do otherwise.

Arne Naess, the Norwegian philosopher who founded the "deep ecology" worldview, believes Kant's distinction between dutiful and beautiful acts is crucial for the well-being of the planet. "When people feel they unselfishly give up or sacrifice their self-interests to show love for nature," warns Naess, "this is a treacherous basis for conservation."[1] Doing right by the Earth should, rather, feel as natural as doing right by our families, our very selves. According to Naess, the way to nurture this mind-set is to expand the boundaries of self into that of Self—the greater self of the planet, with all its creatures and landscapes.

The ecological crisis (which is undeniably genuine) thus demands a deep solution. The will to change must come from within. Only a shift in values can work a lasting shift in laws and institutions and, most important, everyday practices. And those values must emerge from a shift in worldview that is in a fundamental sense religious.

Not long ago it was intellectually fashionable to declare that religion's time had passed. Religious experience—and even more so, religious dogma and institutions—were regarded as drags on human progress. Supernatural belief bound the individual to pre-rational states of consciousness and choked societies with doctrines invented in pre-modern times. Marxists assailed skyward-looking religions for lulling the downtrodden into accepting a wretched existence here on Earth. Nietzsche proclaimed, "God is dead." Meanwhile, secular humanists held a mirror to themselves, turning to humankind and human culture as the only aspects of heaven and Earth worthy of reverence. We ourselves were the beginning and end of all meaning and value.

Smug disregard of the religious impulse has recently fallen out of fashion. Many people now realize that a sense of the sacred need not be based on superstition and supernaturalism. Joseph Campbell, who held that religion was whatever put one "in accord" with the universe, delighted in the mythic metaphors of diverse religious heritages while savaging those who corrupted the metaphor by claiming its material truth.[2] For Huston Smith, religion is that which "gives meaning to the whole."[3] Lawrence Kohlberg judged religion to be that which "affirms life and morality as related to a transcendent or infinite ground or sense of the whole."[4] Theologian James Gustafson puts forth a definition of religion that is as accessible to atheists as to theists and that, moreover, offers possibilities for making peace with the Earth. In Gustafson's view, the religious capacity manifests as "a sense of dependence, of gratitude, obligation, remorse or repentance, and of possibility."[5] Philosopher Loyal Rue defines religion simply as "an integrated understanding of how things are (cosmology) and what things matter (morality)."[6] Note, therefore, that one need not be a theist to be counted among the religious. Loyal Rue is such an example; his religion, which is shaped from a scientific (specifically, evolutionary) understanding of the cosmos, is *religious naturalism.*

The human religious capacity is also being taken seriously today in part because of the work of biologists with impeccable credentials as scientific materialists. These scientists made the astonishing discovery that the religious impulse (for good or ill) may be too deeply rooted to be rooted out.

A Surprise from Sociobiology

Jacques Monod (1919–1976) was a molecular biologist who combined the authority of a Nobel laureate with a passion for philosophy and a gift with words. In his 1971 masterpiece, *Chance and Necessity,* Monod surmised that the capacity for religious experience and the hunger for religious explanation were molded by the same force that shaped our opposable thumbs: natural selection.

Evolution of mental capacities that bolstered group cohesion beyond the innate genetic concern for close relatives would have helped members of larger groups cooperate for the good of all. Scientists writing after Monod recognized that even if loyalty, valor, and the surety of meaning offered by religious belief took a toll on the fitness of warriors who died defending the tribe, such seemingly altruistic acts nevertheless benefited copies of warrior genes carried in the chromosomes of remaining kin. Members of groups made coherent and strong by shared religious conviction thus would have been favored by evolution. "We are the descendants of such men," Monod wrote. "From them we have probably inherited our need for an explanation, the profound disquiet which goads us to search out the meaning of existence. That same disquiet has created all the myths, all the religions, all the philosophies, and science itself."[7]

That this "imperious need" is inborn, Monod continued, that it is now inscribed in the genetic code, "strikes me as beyond doubt." Through the millennia, not only the capacity but also the need for a religious framework entered our very DNA. The drive to find or construct a complete explanation by which to orient ourselves and our goals in the universe is thus innate. Its absence, Monod cautioned, "begets a profound ache within."

Edward O. Wilson took up where Jacques Monod left off. In 1975, with publication of a massive tome titled *Sociobiology*,[8] Wilson founded a new branch of science. Sociobiology draws from the fields of evolutionary biology and population biology to explore the evolutionary roots of all sorts of social behavior in animals—from mating rituals and dominance hierarchies expressed in many species to the very few forms of behavior and emotion that seem to have no analog outside our own kind. Sociobiology thus looks at social behavior from an adaptationist standpoint. How, for example, does an instinct to whistle an alarm call help a prairie dog propagate its genes? How might deception—even self-deception—enhance the evolutionary fitness of an ape?

A few years after publishing *Sociobiology*, Wilson left prairie dogs and chimpanzees behind to focus on the human species. In so doing, he widened his scope to include matters of philosophy and religion. The resultant book, *On Human Nature*, was not a work of science, Wilson cautioned. It was more a "speculative essay"—one that earned its author a Pulitzer Prize. Nevertheless, the science and argumentation Wilson presents on the sociobiology of religion are formidable, going well beyond the groping ideas that fellow biologist Jacques Monod had pioneered.

The predisposition to religious faith is "the most complex and powerful force in the human mind," Wilson conjectured. It is likely "an ineradicable part of human nature."[9] Wilson includes in his list of innate religious qualities the "mythopoeic drive," along with such unsavory items as xenophobia, trophyism, and attraction to charismatic leaders. Those who view religion in a more congenial light might build the list around such things as a sense of wonder, an urge to express gratitude, a capacity for mystical experience, a reverence for whatever is deemed sacred, and a drive to find ultimate meaning in life, suffering, and death.

However one chooses to flesh out the traits of religious capacity, the "collision between scientific materialism and immovable religious faith" noted by Wilson is undeniable. Fanaticisms on both sides of the divide are all too evident. And the casualties—those whose spirits are vacant or fulfilled mostly in shopping malls—greet us on buses, in the work place, in our homes, inside our own skulls. The clash ensues because "our schizophrenic societies progress by knowledge but survive on inspiration derived from the very beliefs which that knowledge erodes."

What to do?

Wilson makes a daring proposal. To reconcile science and religion we must "concede that scientific materialism is itself a mythology defined in the noble sense." According to Wilson, science would remain no less real and right, but it would also be seen as ripe for extension into the realm of myth and meaning and value. Here Wilson parts company with Jacques Monod, who felt that human-centeredness was the only way to cope with existential despair. Not so, counters Wilson. Science offers humankind the grandeur of the "evolutionary epic" for putting ourselves in accord with the universe and urging us on to even greater accomplishments. Note the extension into lyricism and subjectivity implied by the word "epic." The evolutionary epic is not science; it is science extended into the realm of meaning. In such imaginative form, the history of life and the cosmos become the creation story for our time. My story and your story are not just part of the triumphant march of humankind. They are part of the even grander story of the evolutionary stream of life, of planet Earth, and of the universe. Moreover, the grandeur of that story stands firm, even when faith in ourselves and our kind begins to flag.

Wilson thus urges that we satisfy the innate longing for religious grounding with a cultural explanation derived from science. That explanation would be based on the evolutionary epic. It would incline us, Wilson hopes, to dedicate a good portion of our religious zeal to reverence for the vast diversity of life produced by nearly four billion years of struggle and symbiosis on Earth. By way of the evolutionary epic we can redesign our prescriptions for spiritual allurement and atonement, and we can revisit questions of ultimate meaning and value.

All this is possible because the capacity for religious experience and explanation is just that—a capacity. Genes do not tell us how the world came into being. Genes do not determine what we revere. These crucial details are, rather, the workings of the cultural counterpart of genes: memes.

The term *meme* is the brainchild of evolutionary biologist Richard Dawkins. Examples of memes are "tunes, ideas, catch-phrases, clothes, fashions, ways of making pots or of building arches," Dawkins explains. "Just as genes propagate themselves in the gene pool by leaping from body to body via sperms or eggs, so memes propagate themselves in the meme pool by leaping from brain to brain."[10] The term is an artful invention, given its sound resemblance to *gene* (*meme* rhymes with *theme*) and its etymological connection to words such as *mimetic* and the French word *même* (meaning, "same"). The meme of *meme* is now wildly successful. Writers who normally have no truck with evolutionary biology or any other science now find it indispensable. Memes are what give substance to our inchoate capacities for religious feelings. Whether they enter our minds by thoughtful or thoughtless invitation or by indoctrination, particular memes are usually what we judge when we speak of religion in friendly or unfriendly ways.

Overall, even if the mythopoeic drive and other vestiges of the religious are innate, it seems to be a cultural choice whether these are expressed through memes that will impair or enhance our bond with other species and with Earth itself. Both nature-phobic and nature-philic religious memes are evident in the world today. We do have a choice. Moreover, the choice may be genetically slanted in favor of what, in today's vernacular, would be called "green."

Here E. O. Wilson is again a pioneer. He has suggested that a desire to associate with, even to love, living things is not just a cultural choice. In his 1984 book *Biophilia*, Wilson claimed that "to affiliate with life is a deep and complicated process in mental development. To an extent still undervalued in philosophy and religion, our existence depends on this propensity, our spirit is woven from it, hope rises on its currents."[11] Wilson's "biophilia hypothesis" may be highly speculative, possibly unconfirmable; but the idea that love of living things may be genetically imprinted in human nature has been a smashing success. The terminology has caught on, the hypothesis is talked about, and the beginnings of a research program in biophilia are evident.[12]

If humans do have an innate capacity for biophilia, how can it be nurtured? Is biophilia similar to language acquisition—is it a capacity that must be exercised in childhood in order to blossom? Will that capacity wither if we are not exposed to the bounties and delights of living things at an early age, just as the language capacity withers for want of conversation? These and other scientific questions about biophilia are of more than academic interest. The answers will influence the direction of environmental activism.

Biologist Jared Diamond, no less than Wilson a proponent of biodiversity protection, has his doubts about the reality of biophilia. His long association with tribal people in New Guinea, whom he has come to know during his birding expeditions to remote forests, makes him question the rosy view of biophilia. New Guineans are extremely knowledgeable about the biological landscape; hunting and gathering are ways of life. But their treatment of animals can be ghastly. He gives several examples: extracting bones for nose ornaments from the wings of live bats; keeping snared animals fresh by transporting them alive, but preventing escape by breaking their legs; immobilizing cockatoos by bending wings behind their backs and tying the feathers.[13]

Why would Jared Diamond offer as evidence against Wilson's hypothesis the apparent absence of biophilia in one small tribe in New Guinea? What about the love of living things evident in the goldfish ponds of Japanese gardens, in the myriad backyard bird feeders of suburban America, in the altogether new concept of "animal rights"? The answer is that hunter–gatherer cultures are where we must look to assess whether any psychological trait has a genetic basis. The field of evolutionary psychology (a subset of sociobiology in which humans are the focus) builds its hypotheses on the notion that our psyches were honed by the hundreds of thousands of years our lineage spent in the Stone Age. Our brains have had little time to adjust to modern exigencies. If love of living things finds its fullest expression in cultures where the discomforts and outright dangers of the natural world have been removed from everyday life, perhaps biophilia is a cultural emergent.

Unlike biophilia, religious qualities do seem to be common to all cultures. The mythopoeic drive, a sense of the sacred, and other manifestations of the religious are thus even more likely than biophilia to have a strong genetic component. Likelihood is not, however, certitude. Even if a human capacity is found to be universal, there is still a danger in granting it a genetic basis. That all hunter–gatherer cultures know how to fleck flint, for example, does not mean that flint flecking is an inborn capacity. Cultural inheritance from a single moment of innovation during the evolution of genus *Homo*, multiple discovery, or even borrowing—not chromosomes—probably keeps this particular skill going. How, then, could we possibly know that the mythopoeic drive, but not flint flecking, has jumped to the genes?

Maybe it doesn't matter.

Consider: We know in other facets of evolution that certain magnificent traits probably evolved for reasons entirely different from their current uses. They are *exaptations*, not adaptations—serendipitous by-products of selective forces that were sculpting something else.[14] They are after-the-fact discoveries. ("Gee, wouldn't this be great for. . . .") Insect wings, for example, may have started as sails that helped their bearers glide across the surface of a pond. Only later were they coopted for powered flight.[15] The vertebrate jaw probably began as a paired gill arch, exapted into service for clamping the mouth shut when a gulp of water was forced out through the gills. The respiratory function remained even when the structure was exapted again to aid in food getting (simple teeth) and then again to aid in food processing (teeth that could rip) and finally even more sophisticated food processing (teeth that could grind).[16] Similarly, owing to natural selection—perhaps a form of natural selection called sexual selection (involving the often whimsical preferences of the opposite sex)—we seem to have evolved brains with the capacity not just to think useful thoughts but to ruminate about all sorts of strange things. Surely intelligence, in general, was selected for. A by-product of intelligence is that we begin to wonder about the meaning of it all. A satisfactory answer must be found, else we risk falling prey to what Monod called the "profound ache within."

Whether an adaptation, exaptation, or cultural inclination—whether expressed through memes that are reasonable responses to an only partially known reality or just fantasy—the religious urge today is rising like the phoenix. The upsurge in spiritual (and outright magical) tendencies in the former Soviet Union, the attraction of fundamentalist doctrines in the Middle East, and the trend in my own generation in America to head back to church or into a coven is empirical evidence that the religious capacity must be taken seriously.

Varieties of Ecoreligious Experience

Religious memes that can soothe the inevitable human-to-human, group-to-group, and nation-to-nation tensions in an increasingly crowded but interconnected world are surely worth every bit of support each and every human culture can muster. Our kind has been searching for ways to foster goodwill ever since we invented the concept. Umpteen versions have been tried and tested. And the sad fact is that faiths and creeds that promote the most brotherly love may not be so good for sisters. Those that stress compassion may not always be the best at getting the fields plowed and warding off ill-intentioned intruders. Those that promote peace and harmony within the group may be bloody terrors without. More to the point, improving the human-to-human relationship is far beyond my ken. Rather, this book explores new ideas about meaning and value that might improve the human-to-Earth bond. Here lies the mythopoeic and ethical frontier. If we can better our human-to-Earth relations to the point where waters regain their health, eroded hillsides recover, and the terrain is not scavenged for every last burnable and edible, then the human-to-human tensions will be all the fewer. Those of us who choose to focus on earthly concerns are not, therefore, turning our backs on human needs.

In 1990 thirty-two prominent scientists, led by Carl Sagan, put their signatures to a document titled "An Open Letter to the Religious Community." Freeman Dyson, Stephen Jay Gould, Motoo Kimura, Lynn Margulis, Peter Raven, Stephen Schneider, and Victor Weisskopf were among the signatories. The manifesto briefly recounted the story of escalating human impact on the environment. "We are close to committing—many would argue we are already committing—what in religious language is sometimes called 'crimes against creation.'" Problems of such magnitude "must be recognized as having a religious as well as a scientific dimension .... Efforts to safeguard and cherish the environment need to be infused with a vision of the sacred." The thirty-two scientists thus appealed to the world religious community "to commit, in word and deed, and as boldly as is required, to preserve the environment of the Earth."

The appeal was answered by several hundred religious leaders of all major faiths and from around the world. Thus arose a coalition, the Joint Appeal by Religion and Science for the Environment, co-headed by Carl Sagan and James Parks Morton, dean of the Cathedral of St.

John the Divine in New York City. In 1991 the group declared that "the cause of environmental integrity and justice must occupy a position of utmost priority for people of faith." The coalition has since produced a number of aids for religious networking and for environmental education. The Joint Appeal, in turn, spurred the founding of a new organization: the National Religious Partnership for the Environment. This partnership includes national-level groups representing Catholics, Protestants, Jews, and Evangelicals in the United States. It encourages each of the faiths to build an ecological component into its tradition and then makes these products available to priests, rabbis, ministers, and other religious leaders.

In most quarters of Judaism, Christianity, and Islam, it is still heresy to consider trees and frogs and the Earth itself as divine or as manifestations of divinity. Paganism, pantheism, and all forms of nature worship are scorned. Yet it is perfectly acceptable to regard the natural world—the creation (usually now the *evolved* creation)—as a sacred work of divinity. We can then acknowledge our own negligence in failing to serve as good stewards of God's green Earth.

The greening of traditional religious faith is a hugely important component of the ecoreligious movement. But there are other ways, as well, to infuse ecological concern with a vision of the sacred. There are other ways to fill the perhaps innate drive for religious grounding with memes that can serve the Earth community. The ecoreligious revolution is unfolding along five distinct—but not mutually exclusive—paths. These five may be called the way of reform (just discussed), the way of the ancients, the way of transcendence, the way of immersion, and the way of science.

Those who warm to the idea of worshipping Earth directly, rather than through a posited creator, can follow the *way of the ancients*. This path encompasses the nature religions of primary peoples everywhere and the revival of various forms of earth goddess worship. Thus the attraction of Native American and Aboriginal Australian rites of passage and views of the sacred. But those who suffer "the accident of being born to a culture that separates nature and home," as Richard Nelson describes the modern pathos,[17] need not relinquish their own cultural heritage. By digging deeper into the past, we may find ancestral roots more to our liking. For those of European descent, Celtic rituals for marking the quarters and cross-quarters of the calendar are becoming popular. Those looking to add a feminine aspect to the face of the divine can call up the goddess worshipped by Old World agriculturists long before the herders entered into covenant with Yahweh. For the descendants of the African diaspora, the practice of Yoruba is an option.

Several widespread religions that are not "of the Book" don't require much (if any) reform in order to embody an ecospiritual component. Buddhism and Taoism are commonly cited as examples. For these religions, divinity already is in everything; we just don't notice it. Meditational practice inspired by Eastern religions is, however, sometimes viewed as narcissistic by action-oriented environmentalists, especially if the all-consuming goal is higher states of consciousness for oneself. Nevertheless, the *way of transcendence* has a long tradition in which success in communing with "the One" is followed by a return to everyday life, with a new-found compassion for and urge to assist "the Many."[18] Thich Nhat Hanh is a leader on this path. This Buddhist monk has been an inspiration to many in the ecospiritual movement, including Joanna Macy (who, with the Australian John Seed, originated an ecological ritual for consciousness raising called the Council of All Beings).

The *way of immersion* works through direct—even mystical—contact with nature. This form of ecospirituality is available to one and all, whether we have an immense wilderness at our doorstep, a treasured tree in an urban park, or just a chance, for a moment, to float with the clouds through a window. Something deep within us is brought into communion with the mountain, the tree, the cloud. Walt Whitman, Henry David Thoreau, John Muir, and William Wordsworth were exemplars of this faith not long ago. Today Annie Dillard, Diane Ackerman, Barry Lopez, and Richard Nelson are among the growing family of storytellers and bards who offer us their own experiences. Their tales reach into our souls. We learn from these teachers how we mortals, too, might become spiritual beings, if only for a moment, by fully entering into the miracles routine in the world of nature. We can do it on foot or vicariously through *Desert Notes* or *The Moon by Whale Light*. The prophets of the way of immersion can urge us on, but we are all deliciously on our own, for this is a doctrineless path.

Finally, there is the *way of science*. This path draws primarily from the biological sciences—notably, evolutionary biology, conservation biology, ecology, and geophysiology. The more we learn about Earth and life processes, the more we are in awe and the deeper the urge to revere the evolutionary forces that give time a direction and the ecological forces that sustain our planetary home. Evolutionary biology delivers an extraordinary gift: a myth of creation and continuity appropriate for our time. This is the grand sweep of the evolutionary epic. Evolutionary biology and conservation biology together introduce us to our farthest-flung kin, promoting knowledge and valuing of biodiversity throughout the world. We relish life in all its multifarious forms. Ecology, in turn, has a presence in the bioregional movement. Deep reverence is accorded the particular watersheds, nutrient cycles, and biological communities that are the lifeblood of particular human communities. Finally, geophysiology, including Gaia theory, has reworked the biosphere into the most ancient and powerful of all living forms—something so much greater than the human that it can evoke a religious response.

Science is thus ripe for extension into the realm of meaning in ways that can "green" our sense of space and time. We are privileged, through science, to know and witness the immense and fecund journey of life on Earth. Time thus becomes history, and history sacred story. We are privileged, through science, to augment folk wisdom of ecosystems and to begin to learn the physiology of the whole Earth, Gaia. Space thus becomes a cherished place: our bioregional and planetary home.

More broadly, all five paths of ecoreligious experience—reform, ancients, transcendence, immersion, and science—lead to pretty much the same ends. All can engage the mythopoeic and moral capacities of the human psyche in ways that enhance the human–Earth relationship. All can be used to nurture reverence for the natural world—a reverence deep enough to alter the meanings we construct and the values we profess. All can promote beautiful acts of a decidedly green hue.

Each path, moreover, calls out to you and me in different ways. As William James judged in his 1902 masterpiece, *Varieties of Religious Experience*, the human spirit has a range of capacities for religious experience, and the relative strengths of those capacities differ from individual to individual. "Each attitude being a syllable in human nature's total message, it takes the whole of us to spell out the meaning completely."[19] The need for pluralism in cultivating *ecoreligious* experience is just as strong today. It is a pragmatic need. Not all of us can access the mind-set by which we might genuinely follow the way of the ancients. We can't follow the way of reform unless we already adhere to a particular religious faith or unless the prospect of the greening of that religion is sufficient to lure us back. For some, the ways of immersion or transcendence are either impractical or unappealing. And not all will want to make science their beacon—but some familiarity with this path surely can enrich the journey along all the others.

The way of science is the subject of the rest of this book. Whatever your inclination, there should be something helpful for you in the chapters to come. I fear, however, that "the way of science" calls up the sterile image of white lab coats and the stench of formaldehyde. So along the way I will introduce you, through recorded conversations, to the passionate commitment as well as ideas of leaders on this front. The spoken word, as presented here in a variety of conversational extracts, proves to be a marvelous medium for revealing the depth of feeling, the humor, and the humanity of people involved in the greening of worldviews.

Science and Meaning

The way of science has been practiced in the West for half a millennium. Although some scientists have from time to time been appallingly nature-phobic (Francis Bacon, for one), many have been just the opposite. Today in particular, scientists working in any field of biology and, more pointedly, in any of the subdisciplines we shall explore in the next four chapters, are overwhelmingly "green" in their outlooks. And although some scientists have denigrated spirit and religion, others have regarded their work as a way to learn more about a universe that is in some sense sacred. Moreover, proponents of ecospirituality who are not normally associated with science have voiced support for this path. Gary Snyder is one of the most syncretic leaders of ecospirituality in America. He is a founder of the bioregional movement. He is a nature poet, a practicing Buddhist, an acolyte of Native American traditions, and a proponent of a "scientific-spiritual culture." "Let no one be ignorant of the facts of biology and related disciplines," he admonishes, for we must "master the most imaginative extensions of science."[20]

To master the most imaginative extensions of science—this is indeed a worthy goal for those of us who practice the way of science. The attraction of science is both its beauty as a heritage and its prospects for change. School textbooks, unfortunately, sometimes render science as dogmatic as any fundamentalist doctrine. In truth, science is quintessentially open. It is open to revision, to new ideas. Much of what is truly of interest to the philosophical quester is likely to be (and to remain) in the throes of controversy within the scientific community. Respected authorities will favor utterly different theories. For example, there is now no question that biological evolution is the means by which species were created. But which aspects of the pageant of living forms that have graced this planet are due to chance and which to interior or exterior shaping that would have worked the same sculpture in any lump of clay? Choose your authority (Stephen Jay Gould or Stuart Kauffman, in this case) and thereby choose your answer. Choose your answer and thereby choose your worldview.

Science is also open to interpretation. Scientists can tell us what is and what was and perhaps even what will be, but not what it all *means*. For example, Big Bang cosmology has been almost universally accepted within science. Even so, it is up to each of us to decide whether we feel welcome or alien in that sort of universe and what role, if any, a theistic or deistic god or spiritual influence may play in it. Science is thus one of the most important bases for meaning-making in today's world. The meaning drawn out of science by each individual who treads this path is a constructed, but not arbitrary, product of the human imagination. Despite the inherent subjectivity, meaning-making is not mere fabrication. It is a response to, a declaration of relationship with, Earth and the cosmos. To find meaning in the cosmos is no less legitimate than to have an aesthetic response to a landscape. Others may have a different response, but to be fully human is to have a response of some sort.

So far as can be known, we are the only beasts blessed and burdened with a mythopoeic drive. According to the historian of myths Mircea Eliade, we are not *Homo sapiens*; we are *Homo religiosus*.[21] The mythopoeic drive may be inborn or culturally induced, or we may stumble on it in the playful exercise of our minds—noticeable even in young children. Whatever its genesis, the fulfillment of at least a portion of religious capacities may be essential. Our mental health may depend on it, as may the continuation of our species and many others.

The practice of science, as well as its interpretation, is a community effort. No single researcher is obliged to entertain all hypotheses and pass judgment unmarred by emotions or philosophical inclinations. Quite the contrary: Science proceeds by skirmishes between members of conflicting research programs who are often at one another's throats. Submitting a paper for peer review in a controversial field is not for the faint-hearted. Those who have the courage to enter the fray may be empowered by conviction as strong as that of any believer. If their ideas turn out to be wrong, they will leave a legacy of little-cited papers. If they happen to be right, they will win the only kind of immortality most scientists believe possible. Pigheadedness, the self-serving quest for glory, and other all-too-human tendencies held in low esteem (short of theft and fraud) can therefore yield exemplary results when judged and checked by a community of peers. One need not emulate the means to admire the ends. For many of us, even the squeaky clean image of "the scientific method" that we learned in textbooks is unattractive—pallid and dehumanized. *I* may not yearn to be a scientist, but thank goodness some do. Ah, what I can make of the results! In turn, I need not approve of all the technologies drawn out of scientific knowledge in order to applaud the continuing adventure of the scientific quest.

The way of science as a spiritual path thus does not demand allegiance to the methods by which new scientific knowledge is discovered or promoted. Nor does it require one to embrace all the applications of that knowledge. The way of science turns on ideas. It presents to us a picture of the world that we then interpret. Its most exacting demand on a seeker is a good measure of tolerance. The world is real; our scientific ideas about the world are a pretty good approximation of the real (and getting better all the time); but our interpretations of meaning and value are largely constructions. Some interpretations may be more plausible than others. Some may be more useful. Some may provide us with a greater zest for living and acting with commitment.

The way of science is also a bit dangerous for those who don't relish the idea of having to trade in their worldviews from time to time. This is because meaning-making not only emerges in the human realm; it evolves. The way of science may offer the greatest challenge when the news makes us uncomfortable. Overall, those who find their sacred text in nature must be able to live with errata sheets. For example, I remember the psychological loss suffered and the limbo endured when I had to trade in a homeostatic version of Gaia theory a few years ago for a more developmental view of Earth's climate and chemistry. Eventually I found the developmental view to be even more to my liking, but it took a while for me to work up the new interpretations.

Fundamentally, I look to science as a way of extending my experience. Science can't take the place of a hike back Redtail Canyon. But science can carry me to a coral reef in Australia later that evening or to the Andromeda Galaxy. Acquiring knowledge about the carbon cycle is no substitute for a stroll in the woods, but it has many times enhanced that stroll, as I stretch my imagination to follow the breath that I know flows between the trees and me. Fossil bones and traces can't take the place of wide-eyed deer or sun-drenched cottonwoods, but an awareness of the past as revealed by paleontology surrounds me with ghost species that have an eerie presence.

I remember the thrill of encountering a cliff along Puget Sound in which I could read fires and storms in the sediments. I remember the sorrow of driving across Iowa, slicing through a northward wave of migrating Monarchs—my sadness deepened by knowledge (acquired by scientists just twenty years earlier) of how far those butterflies had come, and how far they were determined to go. Ultimately, I have not just my human experience, but a bit of the tanager, the mantis, and the yucca blossom that waits for one nocturnal moth. And I am not limited to just these forty-five years. Through science I have gained memories of animal ancestors that reach back six million centuries. A little more hazily, I recall highlights of my cellular journey long before the first chordate.

With my previous books, I carved out a niche between that of science advocate and helpmeet for the reader. Those books were anthologies of some of the best writing in evolutionary biology that integrated science with meaning, and they sampled the gamut of philosophies.[22] I aim to do much the same here. What I, as an advocate, feel compelled to push in this book has already been said in this first chapter. I will don the hat of missionary once again in the final chapter. Along the way, my green passions will flicker in sync with those whom I engage in conversations. They and I will thereby offer a wealth of ways to use the biological sciences for greening one's deepest worldviews—that is, for nurturing ecoreligious sentiments. Underlying this celebration of diversity is, nevertheless, my commitment to beliefs on which I yield no ground: I believe that a greening of worldviews must occur for the sake of our own species and for that of many co-travelers of the Cenozoic. And I believe that a greening of worldviews must take place at a level deep enough to alter (or even direct) one's religious outlook.

Other than that, everything else is up for grabs. I hope to make the choices so rich and so inviting that by serving more as tour guide than as proselytizer, I will accomplish my greater goals.

2

Science and the Coming of a New Story

"The evolutionary epic is probably the best myth we will ever have." This sentence stopped me in my tracks when I came upon it a decade ago.[1] Since then I have come to agree with its author, Edward O. Wilson—and then some. The evolutionary epic is indeed the best myth our kind has stumbled upon in its tens of thousands of years of weaving stories to explain this fabulous world and our place in it.

Myth as falsehood is not the usage intended. Rather, myth as the grand narrative that provides a people with a placement in time—a meaningful placement that celebrates extraordinary moments of a shared heritage. Those of us who have not only learned but embraced the scientific story of our roots know ourselves to be reworked stardust, biological beings with a multi-billion-year pedigree. We know these facts deeply, and for us the story evoked is as empowering as any tale that has ever come alive in the flames of a fire at the mouth of a cave or in the vaulting echoes of a cathedral. For us, the history of life and of the universe as told by science becomes more than a sequence of strange and arresting events. It becomes our personal and shared story, our creation story, our sacred story.

To call this story the "evolutionary epic," as Wilson has done, is to leap from textbook fact into meaningful extension. The evolutionary epic, in the hands of a scientifically faithful storyteller, is not fiction or fantasy. It is nonfiction in the same way that Wilson's exquisite books—*On Human Nature* and, more recently, *The Diversity of Life*—are nonfiction.[2] The best nonfiction begins with a foundation of fact and then calls on the mind to find a story in it, a compelling and beautifully rendered story.

"I still believe that the epic is the human, narrative mode of thinking with the greatest grandeur." Edward O. Wilson thus confirms his faith in the epic form, during an interview in his office at Harvard University in 1996. "I believe that humanity must have an epic—must have its epics, plural," he continues. "An epic is a grand narrative, usually in poetic form, that utilizes archetypes in explaining a theme that engages all of the nation or all of humanity."

Wilson has a string of accomplishments that, if parceled out more equitably, could easily have brought fame to a half-dozen people. He is one of Earth's foremost champions of biodiversity. He is a world expert on ants, having discovered many new species and even genera. Along with Robert MacArthur, he originated the theory of island biogeography (to be discussed in the next chapter). He is the father of the science of sociobiology. In addition, Wilson has written not one but two Pulitzer Prize–winning books, the second on the unlikely literary topic of ants. His writings have spurred the wide use of terms he enlivened, notably *biodiversity, biophilia, evolutionary biology*, and *the evolutionary epic.*

"To give you an example of how deeply I believe in the epic narrative," Wilson tells me, "I wrote *The Diversity of Life* in epic form. The archetypes include cataclysm, rebirth, the summoning of heroes to lead us out of this worldwide crisis in biodiversity. The book closes with the archetype of the new world discovered. It's written deliberately in epic form. I build a tension in the first chapters: A storm strikes the Amazon [near where he is working], but the forest bounds back. Krakatau explodes; one of the great volcanic eruptions of the century wipes out an archipelago, but the cinders are recolonized. Massive extinction events have occurred in the far more distant past, but they were not enough to destroy the crucible of evolution. Now the new menace, the sixth great extinction crisis, is upon us."

"And in your books and public appearances," I interject, "you are summoning the heroes—calling upon each of us to become a hero in some small way." I am reminded of a public lecture by the ecologist Norman Myers the year before. After a thoroughly depressing assessment of ongoing species extinctions, Myers worked a complete shift of mood in his audience. He predicted that whatever generation takes charge and puts an end to the biodiversity crisis will be viewed as heroes for hundreds, even thousands, of years to come.

"That's exactly right," Wilson affirms. "We can make this into a heroic age."

He continues, "The final paragraph in *Diversity of Life* is in fact what, among the Methodists, is known as the altar call. The altar call is that moment at the end of the sermon when the pastor calls all believers who wish to declare themselves for Jesus or to reaffirm their faith to do so by coming forward, to the altar or to the prayer rail, while hymns are sung. So when I wrote, 'Let us meet on the near side of metaphysics—call it Creation if you wish, call it the products of a billion years of evolution—but let us set those differences aside for the moment and agree that one of the great tasks of the immediate future is the saving of the very richness of life,' when I wrote this final paragraph, I meant it to be an altar call."[3]

Edward O. Wilson believes that science offers humankind not only an awareness of the biodiversity crisis and the tools for saving species but also a story that can charge our very souls to take on the task. To succeed, however, the tellers of this story must draw from the wisdom of the humanities.

"I believe," Wilson says, "that the *inspiriting* of material explanation of the world—what I believe to be a careful and honest discovery of the real world in which we live—the inspiriting of that understanding will depend upon co-opting the archetypes and the great narrative techniques that have emerged instinctively during the course of religion, in its service to humanity during the past pre-scientific generations. That service consisting of three parts: *first*, tribal unification and with it the whole armamentarium of moral reasoning and sacralyzed ethical codes; *second*, rights of passage, which are central to all societies and which form the foundation of much of what culture offers for dealing with the human condition; and *third*, the coming to terms with human mortality.

"There has to be a high, serious style developed for the evolutionary epic. Epic is, after all, a poetic form. It's an art form. This is why we prefer the King James version of the Bible and why Shakespeare resonates so. But we have to use a modern genre of style; it can't be archaic."

He continues, "It will not do to simply write a pedantic or a plainly worded book that tells the facts and the evidence. Because where does that leave you? It's only when you strike the inner chords, the mystic chords of emotion, that you are making it possible to transfer some of the energy and seriousness that define provincial religious thought to a secular form. I don't see the poetry or literary style as just a contrivance to accomplish something—like moving a ton of earth or building a flying machine. I look on it as absolutely essential to the integrity of the human mind. So what we must have is poetry within the scientific, physical worldview. That means we need the humanities, too. The humanities could in effect continue to do their thing, but they would have vastly richer material to work with—grander themes—because the real world, the universe—from black holes to the origin of consciousness—offers far more complex and grander themes than does traditional theology."

By this point I am bursting with enthusiasm. "A telling of this science-based story," I say, "must come out of a depth. It must come out of a depth in which there is no artifice because it is so *real*. It's not the storyteller thinking pragmatically that our culture has a need for the evolutionary epic and that therefore they'll have a go at creating it. It's rather: I'm a believer, you're a believer. We are absolutely moved by this story."

"Right," says Wilson. "And when readers encounter the story, they don't think to themselves, 'Well, if I really need it, I guess this could be a substitute for such and such a core conception from religion.' They *feel* it. And they *care deeply*."

Wilson obviously feels it, and he cares deeply. At the time of our interview he was just beginning to work on the second half of a new book that will explore these very issues. He is a soft-spoken man, but his passion breaks through when, drawing from his own work in sociobiology, he exhorts, "Building the evolutionary epic, telling the story: This is our best way to reanimate the deep emotions that are innate to the human mind, having evolved over thousands of generations in a religious context. The self-assembly of complex systems, the evolutionary process: This is the epic we can create by exploring the material world. And there's so much left to explore. It is of such profound and Olympian magnitude."

Those who would recount the evolutionary epic in a way that can "reanimate the deep emotions innate to the human mind" must be masters of compression. They must select from the voluminous facts just those that can best reveal the story of our coming into being and of the lush diversity surrounding us. A drama of fortune and crisis unfolds. There are turning points, close calls, moments of grace or exceedingly good luck. There are ancestors and heroes galore.

Past bards of the evolutionary epic sometimes presented the story as a quest. The Jesuit mystic and paleontologist Pierre Teilhard de Chardin told the story as a "groping" of life toward a kind of Christic unification, the Omega Point.[4] Others, notably the evolutionary biologist Julian Huxley, wove a tale of enrichment emerging step by step, with no goal in sight, no lure beckoning.[5] Daniel Dennett's apt terminology[6] is helpful for distinguishing the two worldviews. Huxley's epic is assembled by "cranes"—built from the ground by earthly processes. Teilhard's is to some extent boosted by "skyhooks"—the pull of a higher, spiritual force. Huxley's version is preferred by scientists, including Wilson.

Whatever the story, the tellers are challenged to evoke a sense of comfort and belonging without compromising truth. Western culture has been suckled on stories in which good triumphs over evil. A yin–yang balance is therefore not sufficient for many of us. Fortunately, out of the interplay of indifferent physical and biological forces emerges a stellar pageant of galaxies and creatures that transcends the dichotomy. The stream of stars blinking on, blinking off, and the living stream of organisms coming into existence, going out of existence, is beyond judgment of good and evil. It is, rather, magnificent. It is sublime, precious, and exceedingly worthy of reverence.

To be powerful, any telling of the story must include an interpretive meaning, but a meaning best nuanced by a regard for how any such meaning comes about. It is crucial to remember that the first fish to set out across a tideflat in quest of a better pool—or, more likely, in desperate search of water, any water, when its own pool dried up—had no inkling that its effort would ultimately lead to feathered flight and cathedrals. Foresight is foreign to the evolutionary process. Thanks to a big brain, however, our species has the extraordinary gift of hindsight. We can render the history of life—even that of the universe—as a causally connected, meaningful sequence of events, just as we can retell our own personal stories in a meaningful way.

In hindsight, one event prepared the way for something to come. Coincidence, even misfortune, was turned into opportunity. But we must remember that at the moment each transition or preparation was happening, the organisms involved hadn't a clue that anything grander might await their descendants. They were just out there looking for another tidepool. The heroes of the evolutionary epic have all been Forrest Gumps.

For example, I can identify in hindsight the public science lecture (by James Lovelock) that I attended in 1985 as the turning point that lured me away from a profession in energy economics back to my love of science. Had I not attended that lecture, I probably would not be a science writer and editor today. But I did not attend that lecture in order to switch careers. For several years thereafter, I read popular science for the sheer pleasure of it, only later making the leap. And I didn't become a science writer in order to write this book. In hindsight, however, the sequence smacks of predestination. I do tend to think of it that way, too, as one stage preparing for the next. But there is also beauty in recalling my personal story more realistically, in the way that Mary Catherine Bateson suggests in her book *Composing a Life.*[7] We take advantage of slim opportunities, swerve ever so slightly to avoid obstacles, and next thing we know we are on a new life course. Theologian Gordon Kaufman has offered a striking term that reminds us of this fanciful, fluky aspect of the evolutionary epic. He calls the process underlying it all "serendipitous creativity."[8]

How ever one may wish to think of the evolutionary epic or its meaning, it surely sparkles as an adventure tale—"the immense journey" in the words of Loren Eiseley.[9] Awareness of this journey grounds us in an awesome deep time almost beyond our ability to comprehend. Here we are, we humans today, surrounded by a singing and slithering and flitting urgency of existence that is all our kin. Here we are, walking and wheeling and flying across vast landscapes that contain the bones of our ancestors, the shells and carapaces of distant cousins. Here we knowers are, as Wilson puts it, "life become conscious of itself," or as Julian Huxley expressed it, "evolution become conscious of itself."

The meaning of the epic, Wilson insists, "is what we choose to make it mean." If we choose to celebrate this living stream that fashioned itself out of atoms, if we choose to revel in the adventure and to envision ourselves as life, or evolution, waking to an awareness of the breadth and depth of existence, then what science gives us in the way of spiritual fare is extraordinary. That awareness is "one of those few phenomena in the universe as we know it," Wilson tells me, "that remains utterly stunning in a literal sense—it stuns the mind to think of ourselves in this way."

The Epic of Evolution

Any version of the epic of evolution told today must begin not with the origin of the first cell on Earth but with the origin and self-shaping of the universe. During the past several decades the discoveries of astronomers, the musings of theoretical physicists, and the ways in which cosmologists have woven such facts and ideas into stories have been breathtaking. Estimates of the number of galaxies in the universe and of stars in our own Milky Way consistently challenge our brains to work at unaccustomed scales. We must think in terms of billions, even trillions. Black holes and pulsars, photons and quarks: These denizens of the cosmos are dazzling in their strangeness. For the universe story to become *our* story, however, amazement is not enough. We need to find and feel a relationship. We must make a connection, sprout an umbilical cord to the cosmos. What out there can offer us relationship?

I am a biology fanatic; my spirituality is Earth-centered. I rarely contemplate the stars. It takes a comet or a better-than-usual Perseid meteor shower to get me out of bed into the cold night. Only recently have I come to appreciate the pre-life stages of the evolutionary epic—owing in large part to the writings and videos of Brian Swimme, a mathematical cosmologist. Swimme has dedicated his life not to furthering scientific knowledge or to teaching scientific facts, but to creating the epic and telling the story in an imaginative and meaningful way. "The challenge today is to integrate the human with the cosmological," he says in one of his videos, "not by rejecting Copernicus and retreating to a former worldview, but by completing Copernicus and entering a new one. We have lots of scientific knowledge, but abstract knowledge is not enough. What's needed is a transformation of a person's entire being."[10]

Brian Swimme's home base is the California Institute for Integral Studies, where he teaches when he is not a featured speaker at conferences and on the lecture circuit. Spanning several worlds, he can comfortably present a talk at the American Association for the Advancement of Science, share a platform with religious leaders, or present the story of the universe to general audiences. Important for my ears is that the science underpinning the story Swimme tells is impeccable. He doesn't need to tap the fringes to find material ripe for myth-making. Through Swimme, I have come to have a deep feeling for two things beyond my beloved planet: two stars. One is, of course, the Sun. In Swimme's words, we are alive because of "the generosity of the Sun. With each second, the Sun gives itself over to become energy. For two million years, humans have been feasting on the Sun's energy, stored in the form of wheat or corn or reindeer." He continues, "Every child of ours needs to learn a simple truth: she is the energy of the sun. And true human fulfillment is found in a face that shines with that same radiant joy."

Swimme urges us to "enter the solar system" by treating sunset and sunrise as opportunities to experience the rotation of the Earth, "like standing on the back of a great whale." He urges us to initiate our youth by taking them out into the dawn to greet the Sun, while we elders tell the story of this star's gift. He suggests these and other experiential exercises because "if you do not experience the universe directly, it doesn't matter at all what you believe about it."

This is not the way the solar system, photosynthesis, and the food chain were presented to me in school. This is science unabashedly extended into the realm of meaning, and even further into morality: "Human generosity is possible because at the center of the solar system the magnificent stellar generosity pours forth free energy day and night, without stopping, and without complaint, and without the slightest hesitation. This is the way of the universe. And this is the way of life. And this is the way in which each of us finds our destiny." And this is why Brian Swimme has switched careers, from scientist to storyteller.

The other star that Swimme has enticed me to get to know is one that he (with the help of coauthor Thomas Berry) named. Amazingly, until 1992, when Swimme and Berry's book, *The Universe Story,* was published,[11] this star—as vital to us as the Sun—had no name. This is the star that singer and songwriter Joni Mitchell surely meant when she composed the anthem of my generation, *Woodstock*. "We are stardust, we are golden. And we've got to get ourselves back to the garden." We are indeed reworked stardust. Scientists first got an inkling of this shimmering fact in the 1950s. We now know that every element on Earth and in our bodies that is heavier than hydrogen and helium was created in the bowels of a supernova that blew up in this sector of the galaxy some five billion years ago. (Actually, our star system was probably born of the colliding outwash of several exploding stars, but one ancestral star is a powerful image for mythmaking.) Sun and Earth are thus by no means Act I of the evolutionary epic. Something came before. Swimme and Berry call this ancestral supernova *Tiamat*.

Act I: Cosmic Evolution in the Universe

Tiamat was the original divine being in an ancient myth of the Middle East. Tiamat's rending (by her ungrateful son) is what gave rise to the great binary of the cosmos that has impressed virtually all peoples: the split between Heaven and Earth. Half of Tiamat became Heaven, the other half Earth. I first encountered the Tiamat myth in a recent children's book and was appalled by the celebration of violence even a modern telling of the myth evoked. In Swimme's hand, however, the theme becomes (here again) generosity. "Some five billion years after the beginning of time, the star Tiamat emerged in our spiral galaxy. Tiamat knit together wonders in its fiery belly, and then sacrificed itself, carving its body up in a supernova explosion that dispersed this new elemental power in all directions, so that the adventure might deepen."[12]

Today's cosmic storytellers are (in many languages) faced with a choice: it, she, or he. In Swimme's first mention of Tiamat, the ancestral star is portrayed as an it. But later in the book, Tiamat emerges with a personal pronoun—and with a purposefulness integral to myths that we moderns must take either metaphorically or not at all. But metaphor is not a lesser presence; it is not an impoverished way of knowing and believing. Metaphor is essential. Only it can tunnel through the cortex and jostle our emotional core.

Along with Swimme's use of personification and teleology come a surge of science and a wash of poetry:

> In the unbearable pressures of a star, hydrogen is burned into helium, helium is burned into carbon, carbon is burned into oxygen. Anything available as fuel is shoveled into the nuclear blast furnace to stave off gravitational implosion. But after billions of years of striving in this way, Tiamat found herself pressed to the wall, exhausted by the effort, helpless to do anything more to balance the titanic powers in which she had found her way.
>
> When her core had been transformed into iron, she sighed a last time as collapse became inevitable. In a cosmological twinkling, her gravitational potential energy was transformed into a searing explosion, a single week-long flash of brilliance that would catch the attention of every watchful creature in the galaxy. But when the brilliance was over, when Tiamat's journey was finished, the deeper meaning of existence was just beginning to show through.
>
> Out of the spectacular tensions in the stellar core, Tiamat had forged tungsten, copper, and vanadium. She vanished as a star in her grand finale of beauty, but the essence of her creativity went forth in wave after wave of fluorine, astatine, and bromine. Tossed into the night sky with the most extravagant gesture of generosity were cesium, silver, and silicon. . . .
>
> Tiamat had forged calcium, a new presence that would one day support both mastodons and hummingbirds. Tiamat had forged phosphorus, which would one day enable the majestic intelligence of photosynthesis to appear. Tiamat had sculpted oxygen and sulphur, which would one day somersault with joy over the beauty of the Earth. Great destruction, unbearable violence, and out of this Tiamat invented the cosmic novelties of carbon and nitrogen, two astounding powers that would one day sparkle

> as life, as consciousness, as memories of beauty laced into the genetic codings. Tiamat's story—the story of her brilliance, her creativity, her passion, her destruction—is a sacred intensification of the universe's journey. In her story we witness a burst of glory, an amplification of the universe's beauty, and a dangerous and joyful release of power.

This assuredly is not textbook science. And it is not for everyone. But it is for some ones. No single approach to the evolutionary epic is going to work for every person who might be receptive to the scientific story of origins. No single approach will work for every age group, every occasion. Swimme's dramatic rendering of the epic is not likely to be judged suitable for an eighth-grade science class, for example, or for a college course in astronomy. But it is eminently suitable as discussion material for any science or humanities class in any college that hopes to bridge the chasm between the two cultures, between fact and meaning, truth and beauty. It is also fit for presentation in a liberal church, for bedtime reading to children, for a ritual around a solstice—even for staged dramatization to commemorate the fiftieth anniversary of the United Nations (more on that event later).

I admit it took me a while to warm to Swimme's full-scale mythmaking, and longer to be inspired by it. Even the name Tiamat at first turned me off. But then I began to think, well, there's Jupiter and Mercury and Venus and Mars. It wasn't until I created a family ritual of the Tiamat story with two of my young nieces (more on that later, too) that I became an enthusiast.

Brian Swimme's work is science nonfiction, escalated to the level of myth. I asked him about tensions he felt, as a trained scientist, in rendering science in this fashion.

"I am criticized sometimes as being too poetic or too enthusiastic," Swimme responded, "but basically my life's work *is* the poetry of science. I realized this very clearly a few years ago when I was giving a talk at UCLA. It was a great moment for me. I got to meet the scientist who had discovered the world's most ancient form of life. I mean, think of actually finding a fossil cell that's almost four billion years old! But when I asked him what he felt when he first beheld the structure of this preternatural being, he said, 'Nothing at all.' I expressed surprise, and he said, 'You try staring down a microscope at dust particles for sixty hours a week. See what *you* feel.'

"That's when I realized that he and I were involved in an unspoken collaboration of sorts. If his life were dedicated to the difficult task of sifting through the data for the hidden truth of things, mine would be focused on celebrating in poetry the epochal significance of what he and other scientists had discovered."

In his rendering of the evolutionary epic, the Universe Story, Swimme draws attention to one-time-only events. These include, of course, the origin of space, time, and everything—which has unfortunately come to be known as the Big Bang. He and Thomas Berry offer an alternative name: the Primordial Flaring Forth. That's a mouthful, so I have come to prefer the name created by children's book author Philemon Sturges: the Great Radiance.

Another unique event in the epic stressed by Swimme is the formation of galaxies. Stellar formation and stellar death are ongoing, but the galaxies happened only once. No new galaxies are being formed today. Through the most powerful telescopes, we do have the privilege of observing some very young galaxies—but only because they are so far away that the light we receive from them today began its journey billions of years ago. Who knows what those galaxies look like now? Similarly with the stars even in our own galaxy: A casual gaze into the night sky reveals a surprising depth in time as well as a vastness in space. The number of light-years over which our eyes can detect distant starshine is also the amount of time we can experience all at once. This is a far fatter slice of time than we can witness simultaneously by daylight. Consider: The photons from the closest star of the night sky, Alpha Centauri, began their journey four years ago; the photons from the most distant celestial presence discernible through binoculars, the Andromeda Galaxy, set out more than two million years ago. Who knows how the stars that generated those points of light may have changed in the meantime?

Act II: Biological Evolution on Earth

On Earth, unique events abound. They are ripe for mythmaking. Life may or may not have arisen more than once on Earth, but we do know that only cells with right-handed DNA and left-handed amino acids survived (or came after) the early holocausts, when every planet in the solar system, including our own, was sweeping up errant debris. All organisms alive today ultimately owe their existence to a single ancestral Ur-cell. A mid-century evolutionary biologist, Theodosius Dobzhansky, and many others have extolled this "unity in diversity" that a knowledge of life history has brought to us.[13]

In turn, each species is a singularity. With the exception of one laboratory example of a duplicated speciation event in the sunflower genus,[14] speciation is a one-time-only happening. The chance mutations, the chance survival, and the selective forces that shape a new lineage through time are astronomically unlikely to recur. That is why extinction is forever.

Finally, with the exception of clones, each organism is a one-time-only event, a unique cocktail poured from the gene pool of an interbreeding population. Even in the rare cloning occurrences in humans—identical twins—developmental forces in the womb and physical and mental shaping in the early nurturing environment ensure differences in genetic expression and even more so in personalities.

Human awe is sparked not only by singularities. It is kindled as well by multiple occurrences that seem like they ought to be singularities. We do a double-take upon encountering identical twins, for example, or a name in the phone book of another city that matches our own. At the level of species, what astonishes us most is not that nature has produced singularities but that repetition is rampant. This is the phenomenon of convergent, or parallel, evolution. Alister Hardy and his contemporaries extolled the majesty and mystery of convergences in popular books of mid century.[15] With the exception of Richard Dawkins's 1996 book *Climbing Mount Improbable*, convergent evolution has, however, fallen out of fashion as a topic for study and reflection among today's evolutionary biologists. But it deserves resurrection.

Many of the most striking convergences were already noted by Hardy and colleagues. Most famous is the breadth (and detail) of similarities between the marsupial mammals of Australia and the placentals elsewhere. Australia drifted away from the other continents before the placental lineage emerged from the older, marsupial stock. Until humans and dingos came in by canoe and rats and rabbits by ship, the flying bats and a few small rodents (that presumably arrived on driftwood) were the only placental mammals in Australia. All the other Australian mammals gave birth to early-stage fetuses carried to term on milk from nipples sequestered in a pouch on the belly. But show a Tasmanian devil to an Alaskan and you'll be warned to be wary around that vicious wolverine. Produce a photograph of a thylacine (sadly, extinct in just the last few decades) and the animal will be pronounced a wolf—though a wolf with an awfully strange pelt that sports tigerlike stripes. Now tell your Alaskan friend that wolverine and wolf are more closely related to a mouse (or a musk ox, or even a whale) than they are to devil and thylacine, and you will arouse either astonishment or derision.

Convergent evolution is good material for mythmaking. Biological singularities express the whimsy of the cosmos, the chance of mutations, but convergences suggest that there are powerful functional attractors that selection drives toward or powerful developmental pathways that are simply in the nature of the way things are. George Gaylord Simpson, one of the great evolutionary biologists of a half-century ago, presented a striking example of convergence in one of his popular books promoting the evolutionary worldview.[16] Wings, he reported, have developed from forelimbs three distinct times: in the reptilian pterosaurs (including *Pteranodon* and *Pterodactylus*) that were contemporaries of the great dinosaurs, in the birds, and in mammalian bats. We now know that the story of winged convergence is even grander than Simpson's portrayal, because a fourth independent origin has been discovered. The largest of the tropical fruit bats, the so-called flying foxes, hail from a distinct lineage that evolved wings on its own.

A stroll I took through the American Museum of Natural History a year ago turned up a striking example of convergence in vertebrate facial weaponry. Horns on the snout of a stuffed rhinocerous looked remarkably like horns on the fossil snout of a giant brontothere (also known as titantothere), an extinct mammal of a different lineage. And those mammalian facial accoutrements looked uncannily like those on the far more ancient snouts of *Styracosaurus albertensis* and *Centrasaurus apertus*—dinosaurs displayed in the adjoining room.

The most celebrated example of convergence in the animal kingdom is the eye. In fact, the intricate and exquisite eye has long been treated as evidence not of evolution but of divine design. Biologist Richard Dawkins is unmatched as a popular expositor of how easy it would be for the eye to develop step by step, with each step not only functional but also a slight improvement over the previous. He shows how an eye need not emerge fully formed to be useful.[17] The ease of optic evolution is also suggested by the widely held inference that organs for discerning light from dark, for detecting movement, or (at the top end) for discerning shape and pattern have arisen independently in at least forty—and possibly sixty or more—different lineages. Ex-

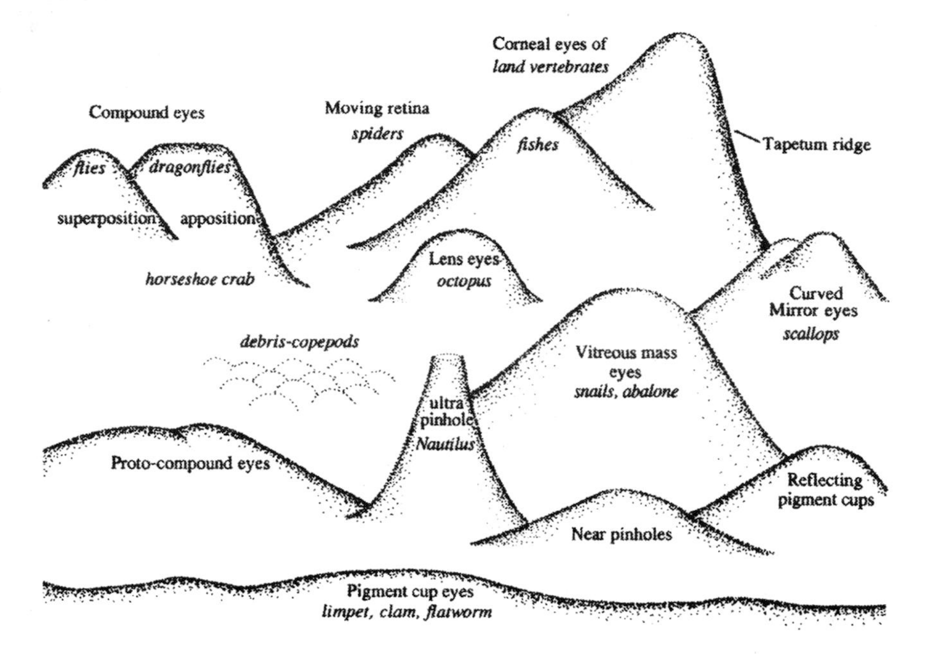

CONVERGENT EVOLUTION OF EYES. In this evolutionary landscape of eye evolution, higher elevations signify higher acuity. Adapted from Richard Dawkins, 1996, *Climbing Mount Improbable* (New York: Norton), p. 195, which was based on information supplied by Michael Land.

changing a glance with an octopus is not all that different from exchanging a glance with another of our own kind. Dragonflies and butterflies, however, have utterly alien visual equipment; we cannot even begin to comprehend their worldview.

The most striking example of convergence in the plant kingdom inhabits the deserts of North America on the one hand and the deserts of Africa on the other. American cacti and African euphorbias are amazingly similar in their drought-adapted bloated stems, spiny armaments, and absence of leaves, but they represent two distinct families. Moving to the algal kingdom, we find a far more fundamental convergence. The red, green, and brown algae photosynthesize by three different kinds of plastids embedded within their cells. These plastids are thought to have been acquired by outright merger (symbiogenesis) with three distinct representatives of the bacterial kingdom on three different occasions.[18] Similarly, the chimera (and hence the bane) of all classification schemes, the lichen group, is now thought to be an amalgam of no fewer than five distinct lineages of fungi that established intimate associations with algae or cyanobacteria within their tissues on five different occasions.[19] Lichens, therefore, are not so much a taxonomic group as a functional guild. The lichen habit is such a powerful adaptation for extreme environments that nature has pronounced this path to be what the Buddhists refer to as "right livelihood." From frozen antarctic granites to just-cooled Hawaiian lavas, lichens are ubiquitous wherever nobody else wants to live. Uncompetitive in a crowd, lichens are nevertheless masters of the near-impossible. Lichens are thus ripe for mythmaking, an Aesop's fable for science-minded parents.

In convergence, distinct lineages solve similar problems in very similar ways, and they make the same creative leap to take advantage of opportunities. But the evolutionary epic is also astonishing in its flush of divergent products. Different lineages often bet on different solutions to the same problem, different innovations for the same opportunity. When arthropods and chordates got their start a half-billion years ago in Cambrian seas, they adopted different ways of supplying bodily support and platforms for muscle attachment. Our ancestors chose an internal skeleton of cartilage and (later) calcium phosphate; the arthropods chose an external carapace largely of chitin—a biochemical relative of cellulose. Crabs therefore sustain the extra burden of having to molt in order to grow, but they are a lot less susceptible to cuts and bruises than we are. When the same two groups later moved out of the sea onto land, they both had to devise ways to extract oxygen from the air. Our ancestors traded gills for lungs; theirs chose to construct for air exchange a body-wide system of pores and tubes that is entirely distinct from the blood circulatory system.

To us it may appear that the terrestrial arthropods—the insects, spiders, scorpions, mites, and millipedes—made a terrible mistake. These creatures rely on passive diffusion of gaseous oxygen throughout a branching system of air tubes, whereas chordates bind molecular oxygen to blood cells and actively pump the liquid through a circular circulatory system. Insects and their kin are therefore far less efficient in delivering the vital oxidant to internal tissues. Terrestrial arthropods cannot afford to dabble in gigantism. No insect stouter than a mouse can make do at today's ambient levels of oxygen. Moreover, even given a billion more years to try, insects probably will never be able to evolve our kind of individual intelligence. The arthropod respiratory system simply cannot promise a big brain a rich and secure supply of oxygen. But insects can show us a thing or two about social intelligence, with their hive minds. Even on the size issue they are not defeated; a swarm of bees or an army of ants is a kind of multicellular superorganism. Consider, too, the ubiquity and sheer numbers of ants and termites around the globe. Who, then, made the wiser choice?

The chordate–arthropod dichotomy offers another example of divergence in the route taken for gaining access to the sky. Pterosaurs, birds, bats, and flying foxes reworked their forelimbs; insects sprouted wings from gill-like appendages.[20] Insects gifted the solar system (maybe even the cosmos, depending on your view of E.T.s) with powered flight long before we chordates even began to try.

Moving to the plant kingdom, we can admire divergent creativity in a wholly different realm. Desert plants, for example, adapt to drought in many ways. They can dispense with leaves and bloat their stems, like the cacti and euphorbias. They can excel in deciduousness, like the mesquite tree of the American southwest, whose boom-and-bust cycle of leaf growth and loss coincides with the rains. Evergreen pines are not so fickle; they have heavily invested in resins to coat their needlelike leaf surfaces and plug up their pores during drought. And then there are the annuals—the lovely, flowering, delicate annuals—that put all their drought resistance into a seed. The annuals have traded away generational longevity in favor of a brief but zestful life. Ah, the tales that can be woven from just the hard-won greenery of a desert!

If Walt Whitman had taken a close look at that annual seed and other seeds encountered on journeys along the open road, he doubtless would have sung of their wayfaring skills. Who says plants are immobile? From the tufted seeds of the cottonwood and dandelion that ride the wind, to the hooks of many a burred interloper that waits for the feel of passing fur; from the explosion of a legume pod, to the come-on of a succulent fruit: These are all ways by which immobile plants mobilize their progeny. And what are showy, sweet-smelling flowers but lures and rewards for favored pollinators?

The list of evolutionary accomplishments is endless. There is the mystery of spider silk, which offers a tensional strength that human technology cannot even begin to match. There are the fish and worms and such that thrive in the ocean benthos, in the darkness at pressures of a thousand atmospheres. There are the petaled and plumed extravagances of sex. There are sensory accomplishments in abundance: the sonar of whales and bats, the ultraviolet vision of bees, the heat-sensing facial pits of vipers, ubiquitous biological clocks. There are fishes armed with electricity, moths that can taste a pheromone miles from its point of release, magnetically tuned and celestially calibrated migrating birds. Knowledge of all these wonders has been made available for human delight only by way of science.

Singularities like the origin of life and the birth of new species; multiplicities like convergent evolution and divergent innovation; evolutionary accomplishments of sensation and execution; generation upon generation of persistence and change—all these stoke our awe. But what is the point of this pageant? Is there a story here?

Science into Story

Stephen Jay Gould[21] and many other biologists believe that the pageant itself is the point. No deeper meaning need be construed. Darwin himself took this perspective. "There is grandeur in this view of life," he wrote in the last paragraph of his great work. "From so simple a beginning endless forms most beautiful and wonderful have been, and are being, evolved."

Before an understanding of evolution by natural selection, the Creation was depicted in hierarchical fashion as "the great chain of being" or "ladder of life"—sometimes capped by God, with angels and humans rungs ahead of the rest of life. With Darwin came a more democratic metaphor: "the tree of life," which has since been reshaped into "the bush of life" by Stephen Jay Gould[22] in order to do away with any vestige of a main trunk and any notion that height is worthier than breadth. Most recently, Kevin Kelly suggested "the thicket of life"[23] to acknowledge the tangle of heritage posed by symbiotic mergings of once-distinct lineages—notably, the birth of the eukaryotic cell. Another apt metaphor is Ursula Goodenough's "sunburst of life." A widely used depiction of the genetic relationships of lineages (shown here in the accompanying figure) does indeed suggest an explosive sunburst of creativity in all directions. But the sunburst metaphor also serves to remind us of how, ultimately, two-thirds of this creativity is fueled. (The domain Archaea derives energy not from the sun but from primordial chemical energy within the depths of the Earth.)

Goodenough, a professor of biology at Washington University in

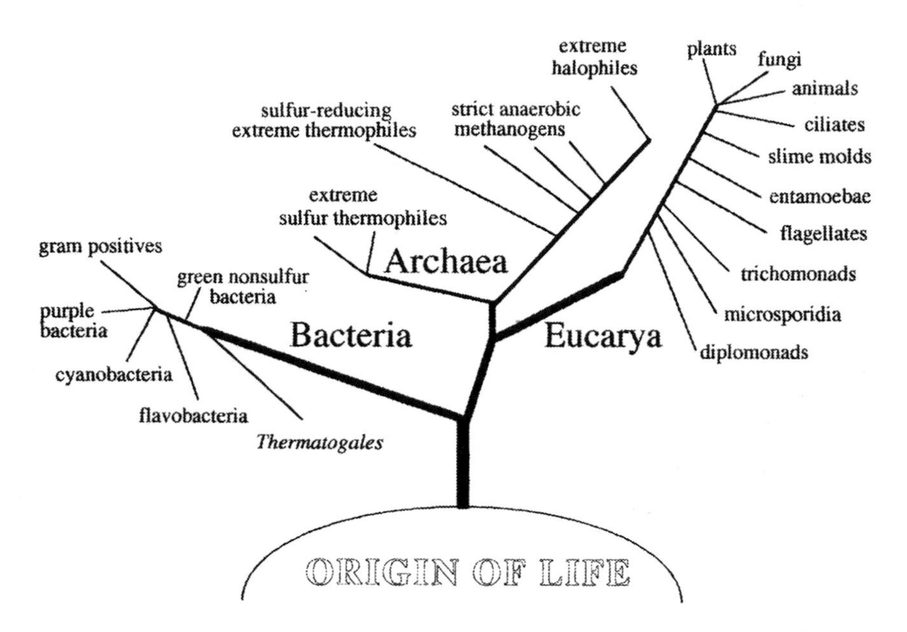

SUNBURST OF EVOLUTION. Adapted from Carl R. Woese, 1994, *Microbiological Reviews* 58: 1–9.

St. Louis, finds in the very first bacterium a locus of meaning and the beginnings of purposefulness in the cosmos. "For me," she writes, "the existence of all this meaning and intent, and my ability to apprehend it, *is* the ultimate meaning and the ultimate value. The continuation of life reaches around, grabs its own tail, and forms a sacred circle that requires no further justification." Goodenough's credo is, quite simply, continuation. "The contemplation of all this continuation, all this connection, all this enormous effort to reach our present level of diversity is for me a deep spiritual resource. I *care* about having it continue. Its continuation is a commandment."[24]

The evolutionary epic is indeed a pageant of seamless continuity and infinite creativity. This by itself is cause for celebration, gratitude, and reverence. Some biologists, however, like to stand back and look at the whole of life splashed across the surface of Earth and roiling in time in order to glimpse something further—something less certain than continuity but perhaps more alluring. Something emerges for these scientists; something enriches within the pageant. A story is born.

Julian Huxley's writings of a half-century ago present a stunning story of evolutionary enrichment, as life pioneers new environments and new ways of making a living. "During the course of evolution in time," Huxley interpreted, "there has been an increase in the control exerted by organisms over their environment, and in their independence with regard to it; there has been an increase in the harmony of the parts of organisms; and there has been an increase in the psychical powers of organisms, an increase of willing, of feeling, and of knowing." This increase has not been universal, Huxley cautioned, as "many organisms have remained stationary [horseshoe crabs, brachiopods, and other 'living fossils', plus an infinitude of bacteria]; many have even regressed [notably, some simplified parasites]. But the *upper level* of these properties of living matter has been continually raised."[25]

Today's evolutionary biologists who search the myriad facts of the history of life for a broad pattern, a story, include John Maynard Smith, Richard Dawkins, Christian de Duve, and Edward O. Wilson. Maynard Smith and coauthor Eörs Szathmáry find that some lineages have become "more complex" with time. The underlying cause? Eight major breakthroughs in the way information is transmitted from one generation to the next.[26] Richard Dawkins tells his version of the story with panache. In his view, the scientific story of life "outclasses even the most haunting of the world's origin myths."[27] Referring to the scientific understanding of our own cells as descendants of communities of bacteria that symbiotically merged into a now-inseparable union, Dawkins finds the idea of "the cell as an enclosed garden of bacteria incomparably more inspiring, exciting, and uplifting than the story of the Garden of Eden." Dawkins scrutinizes the story of life and discerns ten major thresholds, beginning with the birth of the first naked "replicator," thence to the first replicator sheltered in a body of its own making (the phenotype threshold), onward eventually to the many-cells threshold, nervous system threshold, consciousness threshold, language threshold, cooperative technology threshold, radio theshold, and (possibly in the future) the space travel threshold. A big part of the story of life for Dawkins thus resides in the human realm. Similarly, Christian de Duve demarcates seven installments in life's march toward greater complexity: the ages of chemistry, information, protocell, single cell, multicellular organisms, mind, and unknown.[28]

Those who may wish for a more explicitly biophilic rendition of the very same scientific story of life can turn to Edward O. Wilson. As yet, Wilson has given only hints of his view of the "evolutionary epic."[29] Like the other interpreters of the story of evolution, Wilson sees major transitions or thresholds: the origin of life, the origin of eukaryotic organisms (the cellular garden episode that Dawkins celebrates), the Cambrian explosion of trilobite-scale organisms some 540 million years ago, and, yes, the origin of the human mind. The greenness enters with Wilson's original contribution to the story: Biodiversity, he maintains, has generally increased over time. It surged from the origin of life almost four billion years ago, through the billion-year fleshing out of the bacterial kingdom, onward past the florescence of the first amoeba-like eukaryotes to their gathering into the multicellular extravagances of jellyfish and such; thence to the invasion of all nooks of the sea, the land, and finally the air, culminating in the beetle-rich rainforests of today.

The story of life is thus broadly the story of increasing biodiversity, punctuated by five mass extinctions of horrific scale and many lesser extinction episodes. That rise in biodiversity—despite the crises wrought by uninvited visitors from space or by churning geology or (now) by one evolutionary novelty that is all too fit without fitting in—springs from several factors. One, of course, is the dogged continuity of the bacterial realm. Bacteria largely run the nutrient cycles of the surface biosphere, freeing eukaryotic organisms to play with the bounds of complexity. Another factor is the pioneering of fresh environments: the ocean benthos, the land (including the coldest and the driest), and the air. Organisms themselves serve as environments for pioneering parasites and more congenial symbionts.[30] Similarly, but less physically intimate, is the fleshing out of ecological relationships among organisms—predator with prey, flower with pollinator.

I asked Wilson why he viewed the history of life as an epic. He thought for a long while and then launched into a response that transcends biological expertise. "The flat, superficial answer," he began, "is that components of evolution—the great quantum steps—can be shown to represent progress of a sort, and therefore can be construed like a story, replete with crises and emergence into new worlds—the sea, the land, the air—but that's the flat answer. The deep answer has to do with the way the human mind has evolved to work. And that entails archetypes. It entails a compulsion to organize experience in terms of narratives.

"We cannot think without narratives," he continued. "We have an urge to create transcendental narratives, which justify human life on Earth, which justify our tribe, our nation, which empower it by recounting heroic episodes of the kind that bound it together and will bind it again, that will meet any crisis. The adaptive significance of the propensity toward archetypes, epics, is clear." He paused. "You cannot ask why the evolutionary story is an epic without rephrasing it as, Why do we wish to see the evolutionary history of the world as an epic?"

Wilson has not yet explicitly told his version of the evolutionary story in flat-out epic form, and his scientific peers have not even tried. Many have, of course, identified key features of plot and sometimes character, but these scientists still speak primarily as scientists, communicating intellect to intellect. They are not devotedly storytellers, and they are certainly not mythmakers exploiting the power of the epic and its archetypes. Brian Swimme has come the farthest. He revels in mythic language in the biology as well as the astrophysics portions of his *Universe Story.* Not all is sweetly beautiful in his scientifically rigorous rendition of the evolutionary epic, however. Not all is generosity. Here, for example, is Swimme, shortly after introducing the first predator to emerge on Earth (which he names Kronos, the being in ancient Greek mythology who swallowed his own children alive):

> In the future, pale blue skies would shriek with the death terror of pteranodons seized by the quickly stabbing rows of knifelike teeth that lived in Kronos's descendents' mouths. Springboks would learn to eat with their ears ever attentive, lifting their heads silently at the slightest fluff of sound, the doe eyes perfectly still with fright, then exploding in a zigzag escape from the leap of a great cat who had learned its hunting skills from a long line of predators brought forth by the ancestral Kronos. Black eagles soaring with talons outstretched, orcas circling and confusing before shredding the great whale and darkening the seas with its blood, bats whipping through the night to devour thousands of churning insects—none of these would have trembled forth had not Kronos dared to probe this path.[31]

If a mathematical cosmologist can come up with this sort of material, just think what awaits the reading public when a wave of biologically inspired storytellers take a turn at the epic! Here lies a vast frontier for human creativity and meaningful pursuit. And here lies an endeavor crucial for combating our culture's "amythia."

Amythia is a term coined by Loyal Rue, a philosopher and self-described natural historian of culture.[32] It denotes a cultural condition in which individuals are unable to locate the meaning of their experiences and actions in the context of anything greater than themselves, their immediate families, and their own times. An amythic culture may still support remnants, even many remnants, of myths. Symbols and rituals that once were deeply meaningful may still abound, but they have lost much of their power. Overall, there is no common story to elicit pride of group or to enlist passionate action in behalf of the greater good and future generations. Cosmology is diminished, and morality dangles unattached.

"Cosmology without morality is empty," Rue asserts. "Morality without cosmology is blind." In myth, however, we are presented with a compelling weave of both: a unity of what we take to be true and to be good. In myth we thus encounter the integration of cosmology and morality, evoked in a story form that engages the whole brain—feeling as well as thinking.

Many of us, particularly the small subset of Western culture reading this book, are so far removed from any grounding in a cosmic enterprise that we cannot even call up a fantasy of what it would feel like to be bonded with our peers in common pursuit of a grand ideal. Loyal Rue, Brian Swimme, and Edward O. Wilson, however, all wish to initiate a new mythopoeic enterprise, and not just because our culture cries out for it. In their minds, the well-being of other species absolutely depends on it. The ecological crisis demands a mythopoeic solution on a global scale. That solution is commonly held to be some version of what Rue and Wilson call "the evolutionary epic" and what Swimme calls "the universe story."

Why a scientifically based story? "Only a rigorously contemporary myth," claims Rue, "can place our hopes where our energies can make a difference." We have learned from the multiculturalism movement that no tradition should be swept aside, and no tradition should even contemplate going global. Stories of local place and particular histories shall stand and be honored while dreams of hegemony fade. But like a necklace of beads, the string of understanding supported by an awareness of cosmic, biological, and cultural evolution can connect all peoples in a shared story that beckons us to take the rest of earthly life into our intimate circle of concern. As Rue puts it, "There is no reason to think that the value of formulating one's particular story in any way displaces the need for a universal story. In fact, the need for a universal story is amplified by the experience of recovering particular stories. It is true that to know ourselves is to know our particular stories, and to affirm ourselves is to tell these stories. But the more we come to know and to appreciate them the more we become aware of their particularity. And as I discover the limitations of my own story, there is born within me a longing to hear the larger story, everybody's story."

Rue heeded his own advice, given in his 1989 book *Amythia*, by generating his own version of the epic. His book in progress, *Everybody's Story*, is not so unabashedly mythopoeic as Brian Swimme's *Universe Story*, but it does go well beyond science into interpretations of meaning and value. Rue's new book is as much an enterprise in commissioning the story—many different tellings by many different storytellers—as in telling the story itself.

I asked Rue whether everybody's story should be taken as everybody's religion.

"For some, it may become a religion—as it is for me. It tells me in fundamental terms who I am, where I come from, and how I can find fulfillment. For me that's sufficient—more than sufficient, as it moves me to write these books. But there will always be a substantial number for whom everybody's story leaves something out—those for whom the epic of evolution doesn't say all they want to hear. For these individuals, one of the religious traditions will probably also come into play. So let's be clear about this: everybody's story does not pretend to be everybody's religion. The story is fully based on science but goes beyond the facts. So it's no longer science, and it's not just philosophy. Actually, I prefer to use the term 'wisdom tradition' as a label for this story. It will fully serve its purpose when it becomes a bedrock of global wisdom that can inspire us to put an end to the environmental and biodiversity crises."

Brian Swimme concurs, "Like Loyal, I favor 'wisdom tradition.' And for a couple reasons. First, I think what we are creating is more important than a single religion. It is, rather, the comprehensive context for *all* religions—and should be for all other institutions, as well. If what we do ends up being called a religion, it will be boxed off from all the other religions, and that puts it in the same competitive stance that makes for conflicts all around this planet. But if all religions can *draw* from this science-based story, as from a noncompetitive wisdom, then this epic will have a much huger impact. My second reason for prefering 'wisdom tradition' comes from living in California. I would have to check with a sociologist, but I would guess that there are literally hundreds of new religions here in this state. They're being created weekly. I'd really rather not be associated with that enterprise. Overall, I'm happy if the adjectives 'religious' or 'spiritual' are used to describe what we do. I just don't like the nouns!"

Act III: Cultural Evolution in the Human Realm

One more national-level figure in the United States is substantially and passionately involved in moving science into story. This is Thomas Berry. Berry is the eighty-something coauthor (and mentor) of Brian Swimme in *The Universe Story*. Swimme substantially wrote the first two acts that pertain to cosmic evolution in the universe and biological evolution on Earth. Berry wrote the third. Moreover, it was Berry who, a decade earlier, opened Swimme's eyes to the power of this cosmological story.

There is something to be said for the claim that humankind is just another animal, a third species of chimpanzee, better classified as *Pan sapiens* than *Homo sapiens*.[33] There are times when the view that we are just another member of the biosphere—no more worthy than ant or snail—will indeed put us in the right mind-set for doing the right thing. But there are other times when it makes sense to acknowledge our special and unprecedented mental gift, whose dark side is our special and unprecedented capacity to do harm.

This is where Thomas Berry enters the picture. Berry encourages those who read his books and essays[34] and hear his talks to be passionate environmentalists, but without a misanthropic tilt. How? The gift of humanity to the universe, he says, our role in the evolutionary epic, is as knowers and celebrants. "Without the human, the universe could not reflect on and celebrate itself and its numinous origin." Here Berry leaps beyond Wilson and Huxley. For Wilson, remember, humankind is *life* become conscious of itself. For Huxley, humankind is *evolution* become conscious of itself. But for Berry, humankind is *the universe* become conscious of itself.

Mere consciousness is not enough, Berry asserts. Scientific knowledge of the birth of the universe and the pageant of life is not enough. Teaching that knowledge to all of humanity is not enough. The knowledge must spark a celebration. Humans are to tell the Universe Story—what Berry also calls the New Story—in imaginative and empowering ways. Humans are to sing the Universe Story, dance the Universe Story, stage the Universe Story. Berry thus passes through the hubris of scientism like a Star Fleet officer through a wormhole and emerges in a barely explored quadrant of human consciousness where the Two-Cultures conflict does not intrude. In Berry's new world, the humanities and arts have a place of peerless honor.

Berry manages to forge a partnership between the sciences and the humanities, moreover, without a trace of Western or intellectual chauvinism. He acknowledges the truths of the other "wisdom traditions." He acknowledges the wisdom of indigenous peoples, long-time dwellers of the land. He acknowledges the wisdom of women, long oppressed. And he acknowledges the wisdom long expressed in the so-called world religions—including that of his own, Christianity. But unlike Teilhard de Chardin, this Passionist priest and scholar is not proposing a synthesis of science and Christianity. The wisdom of science is so underdeveloped and Berry is so ecumenical that he prefers to leave Christ and even God out of his rendering of the story—though "divine" and "sacred" are indispensable words for him. He prefers to think of himself not as a theologian but as a *geologian*.

Berry has touched many, myself included. His ideas are inspiration for thousands of environmentalists of a spiritual bent. His words have changed the lives and rekindled the motivations of many women and men who share his roots in the Catholic faith but who, like him, had to move beyond roots to connect their spirits with their love of nature and their green commitment. I see his name and clips of his essays and talks more and more in the ecological and ecospiritual magazines I receive. But, not surprisingly, Berry is largely unknown among those primarily committed to a scientific mind-set. Science can do its thing without taking heed of Thomas Berry. Rather, Berry draws from science and then moves into another realm. Science is just one of the feedstocks for the worldview he weaves.

Here is an example of Berry on the role of science:

> We have a new story of the universe. Our own presence to the universe depends on our human identity with the entire cosmic process. In its human expression the universe and the entire range of earthly and heavenly phenomena celebrate themselves and the ultimate mystery of their existence in a special mode of exaltation.
>
> It has taken the entire course of some fourteen billion years for the universe, the Earth, and all its living creatures to attain this mode of presence to itself through our empirical modes of knowing. Such is the culmination of the scientific effort. This endeavor over the past three centuries might be considered among

> the most sustained meditations on the universe carried out by any cultural tradition. Truly the Yoga of the West. Science has given us a new revelatory experience. It is now giving us a new intimacy with the Earth.[35]

Act III of the epic is the emergence of mind. Mind, of course, precedes and parallels human evolution. But it is the human who comes most richly endowed with, and has most exploited, this novelty of creation. Mind leads to culture and thence cultural evolution, which proceeds at a pace that leaves genetic evolution in the dust. For this reason, Richard Dawkins found it fitting twenty years ago to invent a word, *memes*, that gives transmitted thought its due as an evolving presence with enormous consequences for the rest of biology.

Besides speed, another advantage memes have over genes is their Lamarckian style of inheritance. Jackie Joyner-Kersee was genetically endowed with a body that could be shaped into that of an Olympic champion through hard work and persistence. Any children she may bear will have to work just as hard (or even harder) to achieve her level of physical prowess. In contrast, although Brian Swimme has struggled for years to learn science and reflect on its meaning in order to write *The Universe Story*, his children will not have to repeat that struggle. They will receive the meme of the story ready-made and can build upon it in their own adult lives. More likely, it won't be Brian's children who take the next steps. It will be other people's children, perhaps born into households on the other side of the planet. Thomas Berry hasn't even had to invest in children of his own in order to secure a chance of affecting the future of the epic. Memes can jump bodies in ways that genes cannot.

What this means is that in Act III of the evolutionary epic, the heroes need not all be Forrest Gumps. Humans can act with restraint to ensure that the biological pageant continues. We can use foresight to assess the long-term consequences of actions that the reptilian portions of our brains and our own Pleistocene heritage dangerously urge us to pursue. We can even project into the cosmos goals we shape from the meanings we make of the evolutionary epic.

In summary, in Act I things happened, but no thing actually mattered to any thing back in those eons in which the galaxies, the stars, and the elements were coming into being. The cosmos offered nothing then but formless energy and unpatterned atoms. In Act I, for example, the star Tiamat was not striving toward supernovahood or away from it—although a mythic rendering of this event might well slide into teleological language.

In Act II survival and reproduction began to matter. Patterned atoms driven by a flux of energy shaped themselves into the first cell. Life by its very nature is teleological: it has a project—to persist and to beget, which gives rise to the grand pageant of biological beings. But the pageant as a whole may not have any meaning. All meaning is local, forged by the twin goals of survival and propagation held in the memory of each genome.

In Act III, however, the meaning-makers of the cosmos enter. Stunned by the beauty and mystery that surround them, they create stories in an effort to understand who they are, where they came from, and what things matter. Later, stunned by the cosmic adventure revealed to them by science, their own creation, the meaning-makers falter, confused by conflicting stories. Perhaps they will soon reawaken their faith and become busy with new constructions: celebrating the evolutionary epic and conscientiously using memes for the good of the whole Earth community. Life, evolution, the universe can then become exuberantly conscious of themselves. Then everything will matter.

"We may be approaching a critical mass of literature and thought," muses Ed Wilson, as we near the close of our conversation. I have asked whether he thinks that the time may have come for the evolutionary epic to move out into the world where it can make a difference. Julian Huxley wasn't able to accomplish that feat fifty years ago. Neither was Pierre Teilhard de Chardin, though his influence is still felt. "With just enough people concerned, and kneading and pushing," Wilson suggests, "something may finally happen."

A Conversation with Catalysts

In July 1996 the Institute on Religion in an Age of Science (IRAS)[36] held a conference, "The Epic of Evolution," on an island off the coast of Maine. Five participants joined me for several hours of taped conversation. All are catalysts promoting an understanding of the evolutionary epic and its implications. Three were plenary speakers at this conference: Loyal Rue, Ursula Goodenough, and Mary Evelyn Tucker. Loyal (author of *Amythia*) and Ursula (president of IRAS and originator of the credo of continuity and the sunburst metaphor) cochaired the week-long meeting. One participant (Brian Swimme, coauthor of *The Universe Story*) delivered a reflective talk each morning. I was in charge of a discussion session one afternoon on "the evolutionary epic for a greener future." Two participants have doctorates in the sciences (Ursula in cell biology, Brian in mathematical cosmology). Three have doctorates in the humanities (Loyal in philosophy, Mary Evelyn and John Grim in history of religions). Everybody except me is on the faculty of a university or institute, and each of us has written or edited a book or two.[37]

We all knew of one another before this event, but for some of us it was our first face-to-face meeting. Loyal and Ursula have long worked together on IRAS events and as co-organizers of sessions on religion and science for annual meetings of the American Association for the Advancement of Science—for which Brian has twice been a featured speaker. John, Mary Evelyn, and Brian are intimately connected through their studies with Thomas Berry. John began his Ph.D. work with Thomas in 1968; Mary Evelyn began a masters program with Thomas in 1974; Brian was re-energized by the year he spent with Thomas in 1982, after he had become disillusioned with academia. I connected with John and Mary Evelyn several years ago through Thomas, as well. Now we four serve on the board of the American Teilhard Association, with John as president, Mary Evelyn as vice president, and Thomas as president emeritus. I included an essay of Brian's in my previous anthology, and about a year before this conversation took place, I began a lively correspondence and e-mail exchange with Loyal, since coming upon his book *Amythia.*

The conversation begins with each of us telling a snippet of our personal stories: how we came upon the evolutionary epic. Loyal Rue begins. "For me it all started with ecology. I can't date it really, but it was probably '65. A friend and I were driving along in a car and he was telling me about plankton. I was astonished to learn that all of life could be contingent on something I'd never heard of. I asked him where he had learned this, and he said he was taking a course in ecology. I had never heard of ecology. So I went along to his class and listened in on some lectures. Fascinated, I started buying books on ecology, as a concern for the Earth was coming together for me then.

"I grew up Lutheran," Loyal continues. "My father was a Lutheran pastor. My mother was very pious and careful about transmitting the tradition. But when I began to learn about ecology and the environmental crisis, I could see that Christianity had nothing to say to it, nothing at all to say to the greatest challenge to the Earth." Loyal sums up his story: "The ecological morality was there very early on for me, but it took a long, long time to connect it with a cosmology. Once the evolutionary epic did begin to take hold, I got my mooring. Finally, I could justify my beliefs."

"Your story is very similar to mine," I say. "I've been an environmentalist for a long time, since college. But until a few years ago, I never connected that commitment with anything. I lacked a mooring for it. But now I have the evolutionary epic. Now I have a basis, a really strong and solid basis, for that commitment. So even if I'm not sure how I feel on this issue or that, I don't worry, because the morality is not stranded out there on its own.

"I had heard of evolution early on," I continue, "ninth grade biology and surely somewhere in my college training as a zoology major. But the history of life on Earth did not become an epic, a meaningful worldview, for me until many years later. That happened when I stumbled upon the mid-century writings of Julian Huxley. I went through every book of his I could find in the library. And that then led me to so many other beautifully written and philosophically engaging books by other scientists—the kind of writing I had been utterly unaware of during my textbook days. As a science hobbyist rather than a student, I made the whole library my domain, roaming wherever my interests carried me. I remember reading Darwin's *Origin of Species* over the course of three days, outdoors, with fuchshia ablaze and hummingbirds all around. It was a pivotal experience for me. So was reading E. O. Wilson's *On Human Nature* and your *Amythia*, Loyal. Another hugely significant event was coming upon the Teilhard Association just a few years ago, where I met Thomas Berry and Mary Evelyn and John. I mean Thomas was out there *doing* it—being moved by the story and spreading the word."

John Grim has the deepest roots with Thomas Berry, having begun his doctoral studies under Thomas at Fordham University in New York City in 1968. And it was through Thomas that he met his future wife, Mary Evelyn Tucker. John and Mary Evelyn both grew up Catholic; they therefore share the religious heritage of Thomas, a Passionist priest. But they now also embody the two additional worldviews that have most deeply shaped Thomas's thought: those of indigenous peoples and Chinese Confucianism. John is an expert in indigenous cosmologies and Mary Evelyn is a Confucian scholar, but they are more than experts. Like their mentor, they have embraced these traditions within their own worldviews.

"I had an experience when I was in high school," John begins. "I was working on a survey crew one summer on a remote job in the forests of western Minnesota, near where I grew up in North Dakota. A chain man and I had two days to go between the known markers. It was an idyllic moment in my life, so beautiful and blue. At some point I bent down at the end of the tape and when I was coming up the surrounding field fell away. It just fell away. I've since seen a Georgia O'Keeffe painting with the whole canvas black, except it browns toward the top and there's a thin, light pigment on top. My experience was very much like that. And as I stood looking at this, a voice said, 'When your life goes to pieces, you can come and stand here.' That experience with the Earth became a basic impulse in my life." Thomas Berry picked up on John's interest in visions and the natural world and directed him onto the path of studying Native American cosmologies.

"The Native American influence on Thomas is clear," Mary Evelyn offers, "the sensitivity to the land and ritual and symbol and so forth. The Confucian aspect is there, too, in his extraordinary sense of responsibility to the social and political order and a commitment to education. Confucianism has a deep cosmological sensibility and it promotes an engaged system of self-cultivation, all for the transformation of the larger order." Mary Evelyn now begins her story. "I read Teilhard de Chardin in high school, but I didn't understand him until I was in Japan and was sent some of Thomas's writings. I had done a master's in English, realized there weren't going to be any jobs in it, and got an opportunity to go to Japan. Japan was a key experience for me in many ways. I had been going through a deeply existentialist period; being in the sixties was like wandering around in a suffering, myth-of-Sisyphus kind of universe. All of a sudden, enountering this other culture that was so different just blew me apart, in good ways but also very painful ways. I studied a lot of Buddhism, meditation. But when I came back to the U.S. I moved into Confucianism because of its commitment to the social, political, and educational arenas."

Brian Swimme tells his story. "When I was a child I was nature-based. I was always interested in the stars. Like Mary Evelyn, I read Teilhard de Chardin in high school. I went to a Jesuit high school, and because Teilhard was a Jesuit his thought came to me as representing the truth, that this is really what's going on. In college and graduate school I just kept studying; I wanted to know more and more about the universe. Teilhard had strongly activated my interest because he made the meaning of scientific understanding explicit. It was right there. In graduate school I gravitated to the writings of Alfred North Whitehead, as Teilhard didn't have that kind of careful metaphysics. And then it was C. G. Jung and Joseph Campbell. I was always reading those kinds of books, along with getting through graduate school. I just refused to become a person who did nothing but mathematics. I made time for these interests by not going to class, but studying at home. Other students began to call me Neutrino Man because I'd show up and get my mail and then disappear again for a week."

Brian got his degree and began to teach at the University of Puget Sound in Tacoma, Washington. He chronicled his disillusionment with academia in the prologue to his first book, *The Universe Is a Green Dragon,* which is dedicated to Thomas Berry. "I was supposed to introduce students to the universe, the *universe,* but I was not to speak of meaning. Doesn't that seem like a strange assignment?"[38]

"Connecting with John and Mary Evelyn's story," Brian continues, "in '82 I was just this miserable former professor. But then I read an article by Thomas, and I just thought, Wow! So I showed up on

Thomas's doorstep at his Riverdale Center for Religious Studies north of New York City. I went there just to study with him, and that was without question the most important year of my whole education."

"Thomas just bloomed," adds John. "His conversations with Brian drew him deeper into the transition of religion and science. He was already reading the accessible works in evolutionary science, but having Brian there was tremendously helpful."

"So it was really a mutual relationship," I surmise. "Brian brought much of the science to Thomas."

"Definitely," says Mary Evelyn. "The coming of you, Brian, into Thomas's life sparked a florescence. It was surely one of the great moments of his life. He began to have an invigorated sense of hope about the possibility of creating the context for more mutually enhancing human–Earth relations. The collaboration really took off around *The Universe Story* manuscript."

Ursula Goodenough tells the final story. "How I discovered the evolutionary epic is probably less interesting than for any of you because I'm a biologist, so it's a little hard to avoid. I never had a science course until I was in college, however, where I began as a literature major. Barnard had a requirement for science, so I took a zoology course. It was taught by John Moore, who is an absolutely wonderful teacher. Inspired, I went on to take most of the other biology courses. One thing led to another, and I wound up as a card-carrying cell biologist. But still, I wasn't really doing evolutionary work until quite recently, and now that's all I think about. As for the environmental focus, that came to me when it came to all of us—with the garbarge barges, the ecological movement. I was very involved with nuclear issues, even in the early sixties, demonstrating against strontium in milk."

A connection between science and religion has never been foreign to Ursula. She continues, "My father was a historian of religion and a founding member of IRAS. I first came to an IRAS conference on this island in 1987, the year I met Loyal. I discovered this very bizarre group of people wandering around talking about the interface of science and religion. Gradually I found myself getting more and more attracted by this kind of thinking, even while I was trying to hold down a lab and do a straight job. So it became, probably more than for the rest of you, something where I've had to feel a little schizy, because doing this stuff is regarded by my colleagues as off the wall and not the way I am supposed to be spending my time."

Until recently, Ursula—president of IRAS for four years—kept two résumés. One deleted all mention of IRAS and related publications. But when she was elected president of the American Society for Cell Biology, she decided her scientific credentials could withstand contamination by the science-and-religion connection.

Our stories told, the discussion is now open to wherever any of us may wish to carry it. Loyal takes the initiative. "Right now I'm convinced that what we have to do, all of us, is connect up this evolutionary cosmology with a good, strong ecological morality. That's our task. That's what it's all about."

"I have a concern," Mary Evelyn responds. "I think we have to be cautious about presenting the evolutionary epic as *the* story, even as *the* story of cosmic evolution. That makes it unific, and there is no getting around the fact that this story comes out of a particular cultural tradition. We have to be exceedingly attentive to diversity issues here."

"Thomas [Berry] has a nice, inclusive way of putting it," Brian offers. "He says, "This is *a* story of *the* story."

A murmur of agreement fills the room; then Ursula speaks. "One way to put it together is to acknowledge that the religious traditions have several very important features. One feature is that because these traditions are historical, each gives a particular set of humans grounding in history, in time. We as individual peoples respond to our own particular histories. In promoting a shared cosmology, a shared evolutionary epic, there is no reason to scuttle those roots, those histories. It is counter-evolutionary to scuttle our stories; we need all the stories we can get. To live within our own human cultural stories is in no way contradictory to adopting the epic as the shared global story."

"We must remember that cosmological ethics have been developed in many traditions," says John, referring to ethical systems that encourage right relations for human beings with other creatures, the Earth, the universe—not simply right relations with other humans and with the divine. "Buddhism, for example, has developed plural cosmological ethics. Indigenous traditions have cosmological ethics, but they do not have evolutionary ethics. Evolutionary ethics is, rather, a novel form of cosmological ethics."

Brian agrees. "To refer to the evolutionary epic and the imperatives we derive from it in exactly that way would be, in my view, very helpful."

"The whole point," says Loyal, "is that it's anti-evolutionary to say, 'Let's wipe out everything and start anew.' That would be like saying, 'Let's wipe out the genome and get a better one'—a religious mass extinction."

"Cultural evolution has created its own products," Ursula continues. "These religions are to be just as cherished as biological diversity. There's nothing about religion that is anything other than the product of biology. It happens to be neuronal biology, but it's biology. It's great stuff. It provides niches and it provides meaning and it doesn't need to be scuttled in any way, from my perspective, except to the extent that a tradition may say that the evolutionary epic is outright wrong."

"Given all that," says Loyal, "What's important now is to find ways that we can all come to share a cosmology, whatever our tradition."

"There's more to it, I think, than just tolerant pluralism and a willingness to make room for what is deeply embedded," Mary Evelyn asserts. "What some of these traditions can offer to this process is extraordinary. Here's the excitement. We're at this intersection of the scientific cosmological myth and cosmologies of traditions that have empowered people for thousands of years. We realize that fundamentalism is one response to interaction with the scientific cosmology. Okay. It's a process. But there may be something even in that interaction that might lead to some kind of synthesis. Apart from that, what can really be drawn from those traditions, what we're looking for, is the creativity. These traditions have accordioned out to us extraordinary patterns of creative interactions with people, with nature, with the planetary possibilities."

Mary Evelyn continues, "How many times this week have we talked about creativity, the fecund dimension, the changing nature of things? Now consider: The fundamental meditative orientation of the oldest continuing civilization on the planet—China—is exactly on the issue of creative change, on transformation. Confucianism and Taoism—these traditions have a great deal to offer *us* in understanding certain aspects of the deeper meaning of the evolutionary epic. They are helping people to reflect on the nature of change in the universe and how humans can harmonize with patterns of change."

The discussion now shifts to an issue raised this morning during the audience response to Ursula's plenary talk on the biological portion of the evolutionary epic. Loyal introduces the topic. "I really want to spend some time here sifting through the 'sinking heart' problem. Remember when the woman . . ."

"Mary Coelho," Mary Evelyn and I say in unison, as we both know her through the American Teilhard Association.

"Okay, remember when Mary Coelho got up and said, with great passion and great sadness, that the sorts of things she was hearing in

Ursula's talk made her heart sink. What about that? If the Dawkins idea is right, if the continuation of the germ line is at the base of it all, and these bodies, these minds are just stuff to keep the genes going, then that speaks a very astringent message to those who come to the story with some sort of hope and expectation. In my view, any story that promotes a sinking heart is a nonstarter."

I jump in. "What Ursula chose to emphasize in her 55-minute chance to teach the whole biological portion of this epic was largely the mechanics of replication and the taxonomies of diversity and the idea that even the first bacterium possessed a sense of awareness and a capability for response. Important stuff. But the talk was not about the grandeur of the pageant that emerges out of those things. Her talk was not about the emergence of vision, of powered flight, of colorful flowers and their coevolved pollinators, about parental care and sacrifice in all sorts of lineages. As I recall, Ursula did conclude her talk in a very inspiring way, but it may have been too little too late. Overall, people just didn't come away with a sense of the grandeur. Ursula did not highlight the kinds of things that, for example, Brian chooses to emphasize and that draw readers to his books. We need to understand both, of course. I think what Mary Coelho was saying was that we can handle hearing about the less-than-lovely mechanics, we'll take that base part of the story, so long as we get an earful of the rest of the story too—the things that resonate."

"I have an example on this point," says Brian. "Steven Weinberg, one of my heroes, is one of the cosmologists who have really contributed to an understanding of the early universe. His conclusion is that the more you come to know about the universe, the more pointless it seems. I think he's great, but the idea that you can evaluate the universe from the perspective of the first three minutes is madness. It is nineteenth-century physics all over again. There is no adequate way to interpret the story of the universe from anything less than the whole story. It doesn't matter what part you pick; it's going to be inadequate to interpret the whole."

While Mary Evelyn and John and I are sputtering grateful concurrence, Loyal steps in. "Let me bring in the sinking heart from a slightly different perspective. I think that one thing any story has to do, and something that the Christian tradition certainly has done, is to address itself to suffering and to provide a context in which suffering finds meaning. A lot of people will say, 'If I don't get to live forever, then what the hell is this all about?' Or, as many attendees learned just two days ago, 'If the sun is going to blink out eventually, then what's the point?' It's really Dostoevsky's idea that if God is dead, then anything goes; if the self is not eternal, then why bother getting it into tomorrow? That's only part of the much larger question of bringing meaning out of suffering. That's what the cross does; that's what the resurrection does: life out of death. You get meaning out of suffering."

"Similarly," offers John, "Siddhartha Gautama brought meaning out of suffering in the southern and east Asian context. Suffering was not seen as something just to be dealt with; it was the whole metaphysical basis of that religion. No self was posited to be saved from out of suffering."

"You deconstruct the self," Loyal interjects.

"No," John counters, "There is no self to deconstruct."

"Okay. You deconstruct the *illusion* of self," Loyal continues. "That's what people come to a story with. They come to the story with a life—some with great urgency and great need."

"We're talking in the abstract now about what a person needs." I attempt to nudge the conversation. "But perhaps it makes sense first to make sure we all understand why we here, each one of us, has so exuberantly embraced the evolutionary epic. Why do we ourselves find it compelling? How does it provide us with solace? How does it answer for us personally the great, perennial questions? What does it give us, and why are we moved to share the good news, to proselytize?"

"Could I begin?" The floor is now Brian's. "I really have difficulty with the word *proselytize*. Other words, too: *minister, prophet*. When I hear those kinds of words, my heart sinks. Maybe it's just my private issue, but it sounds like self-righteousness, and I worry about the dangers of hubris. That kind of language hooks us into a form of consciousness we're trying to get out of. In contrast, I like to think of what I'm doing as sharing what I'm thrilled about. What am I thrilled about? I'm thrilled about participating in the excitement of what we're doing. I want to be a participant, I want to be an activator, I want to *charge* people. If there's an erotic dimension to this, it's something like that. But when I hear this effort cast in terms of proselytizing and so forth, then that participatory, erotic dimension disappears for me. It just disappears."

"I look to the model I've been most involved in," says Ursula. "And that's the women's movement. In the women's movement, we were all doing it. We were all forming our CR [consciousness raising] groups, writing our letters."

"The feminist orientation is something I feel much closer to," says Brian. "It's mutual."

Mary Evelyn brings the group back to the question that had all but vanished, owing to my overzealousness. "What does excite us about the story, the process, then? Well, what excites me is that the story can't be told without a lot of disciplines, without a lot of different viewpoints, and that excites me enormously. We are in a learning process that accelerates exponentially. This story, this quest, is the catalyst. Our consciousness is beginning to play on a whole different realm of self-awareness and understanding because of it."

"We all have to be involved," says Ursula, "because there isn't going to be an emergent canon. Rather, there's going to be an emergent collection of responses."

Mary Evelyn breaks through the collective affirmations. "And what's neat is that the dialectic is between ourselves and then back to the process, back to the Earth. There's the view, the observation, and then the sharing. The constant dialectic of it is tremendously energizing. It's the sense of discovery and evocation that allows one to enter into the process with the awe but also the action."

"Yes," Brian draws out the word with deep satisfaction. "For me it's a sense of mutual evocation. Every time something is said about the story that strikes me, it's like I come alive in a new way."

"Okay, we've come through that one," Loyal says; then he moves on. "If this is a story that can address the human condition, then it has to offer two things: a therapy and a politics. That is to say, it has to speak to our concerns for personal wholeness and social coherence. That's all we're about, really: therapy and politics. Is there a therapy and is there a politics in this story? I'm very clear about the politics. The politics is what got me into it: green politics, creating the conditions for safeguarding the Earth. But the therapy part—that isn't quite so clear to me."

"Therapy means dealing with sinking hearts," offers Ursula.

"That's right," Loyal continues. "Therapy means creating the conditions for personal wholeness. And politics means creating the conditions for social coherence. Now I think they both have to come out of the same narrative, that they need to be integrated."

"So, you're clear on the politics but not so clear on the therapy, Loyal. Let's explore that," I suggest. "Remember my concern about the passage in your latest book where you call yourself a nihilistic biophiliac? Let's talk about this."

Silence. And then Ursula jumps in to rescue her friend. "Along these lines, I would love to tell a story. My first-born, Jessica, at about the age of three asked what happens when you die. My husband and I gave her a response about going back to the universe, about how we're all connected on this great continuum—we did all that. The next day he and I wondered, however, how a three-year-old could possibly deal with all that. We decided we should give Jessica the most well-received traditional version of an answer, too, so we took her to the best Sunday school we could find in a Christian church. About three years later, she and her younger brother, Thomas, were walking in the woods together—my husband and I were behind—and Thomas asked Jessica, 'What happens when something dies?' After a slight pause Jessica began, 'Well, there's this great continuity of organisms in the universe. . . .' She gave him the whole bit. That was the moment I realized that this epic had potential."

Amid the laughter, Ursula's enormous smile suddenly vanishes. She continues, "My own story in that regard did not have so happy an ending. And this goes back to the issue of the sinking heart." She pauses. "I got a job in third grade accompanying a first grader to school. It turned out he was Catholic, and he started telling me about heaven and about angels and how you get to be with Jesus when you die. He gave me the whole thing. A few months later, talking with my father, this topic somehow came up. And here for me is one of those memories that I will never forget. I perked up and said, 'Oh, it's easy.' And I went on about Jesus and heaven and so forth. My father looked at me like he couldn't believe what I had just done. He said, 'You know, Ursula, that's not the way it is. When you die, you die. It's just over; that's it.' Well, I ran upstairs and I cried and cried. But at the same time I knew that what he said was true. The Jesus thing collapsed. I knew I had to come to grips with dying. But for a long time, a very long time, I didn't have a story in which to satisfactorily place the tragedy of death. I was amythic, in Loyal's sense of the term, until the encounter with this story, this epic of evolution, in my work."

John speaks gently, "Listening to Ursula talk, it's as if in retrospect the story emerges in certain moments, but it becomes tied to a whole range of experiences that are difficult to untangle."

I concur. "It's like once you get it, the new perspective reaches back into everything. All the little pieces of our selves start connecting, once we pick up on the story."

"In Loyal's terms," Brian adds, "it becomes a hermeneutic for interpreting the cultural canon or ourselves. You do a retelling of your life story according to this new hermeneutic."

"I'll give you an example," says Loyal. "Here's my retelling: I was brought to a rejection of my tradition during my college years. My encounter with philosophy taught me that Christianity in the terms it was given to me was intellectually implausible, and after I found out about the plankton, and Christianity's refusal to address that issue, I decided that my tradition was also irrelevant. So I was looking for some way of grounding an imperative that I knew was there. I felt that there was an imperative to save life, save the project. Christianity wasn't doing it. So there was this moral imperative just floating loose—nothing worse than a morality that floats loose. I was in search of a cosmology. I eventually found it in learning more about biology and physics. And that meant that my moral commitment finally had a foundation. All of a sudden, whoomp! It goes down like a taproot. Now it's grounded in the cosmos."

"Any epiphanies?" I ask.

"Lutherans don't have epiphanies." Loyal continues through the laughter, "We try our best to understand these people who from our point of view have to be manic-depressive. Lutherans are emotional flatliners." He waits for the guffaws to subside. "Seriously, I do look for hints of self-understanding in all of this. And I have found it, I think. This story explains to me where I came from and all the rest of it. From the standpoint of the flesh and subjectivity, it does have a bad message. But I've recovered from that. That was a sinking heart experience, but I've recovered. And I've sort of talked myself out of wanting subjective immortality—reasons why I don't want there to be a heaven."

"We'd get bored with ourselves," Brian offers.

Loyal continues, "It's a good thing there isn't a heaven because we'd be . . ."

"We'd be so tired of each other's jokes!" Ursula cuts in.

"We'd have to know what our kids are doing," Loyal insists. "Nobody really wants to know what their kids are doing. Seriously, I don't know what the therapy is for me. Except that it tells me where I am; it tells me the dynamics that brought this kind of experience into reality. Up out of molecules into thought. It tells me how all that emergence happened. The nihilism part is the recognition that my self doesn't matter really, ultimately. And from the universe's standpoint, life as a whole probably doesn't matter either. If you get way out there in the universe, I'm not so sure life is the point of it all. But, hey, terrestrially, terrestrially, on this Earth, I'm for life!"

Ursula is first to penetrate the laughter. "I remember when I really absorbed the fact that the sun is going to blink out about five billion years from now. That understanding, folks, came to me just a few years ago. To think that only five billion years are left for this glorious continuity of life, when five billion years is sort of my entire palette. Do this one more time and then it's over! I had a sinking heart."

Brian shifts the conversation, "I think it requires a good deal of time that we had to live through and think about to arrive at this synthesis. All this was available to me in a certain sense early on—reading Teilhard and being in the sciences—but it just takes so long in our experience to arrive here. The evolution of the stars and galaxies and so forth—all this was fascinating by itself. But I didn't think of the story as having potential for really orienting the world until I read Thomas Berry. Here's another big moment for me: I remember being with Thomas when he said to me, 'Do you see the power of this? Do you see what this means for our society?' And I didn't; I really didn't at the time. The quest was still on just the personal level for me, personal wholeness and understanding. But eventually, Thomas helped me see that this story represents a massive new cosmology and that it can serve as the fundamental organizing myth of our culture. But, you see? It took a long time. So I think it's very interesting we're all sort of in the same age group here. Whether you come at it from politics or environmentalism, I just think it's really interesting. I mean, why isn't there somebody here who's 27?"

"I think it's the sixties," says Ursula. "When we were 27, we were flipping out on what was happening!"

"No question about that," Brian continues. "But I'm saying also that the synthesis we're talking about is formed now in a way that is comprehensive. That's why we can have this conference. But that wasn't true five years ago."

"We know about it because of happenstance or whatever," I add. "And so we're into it. But it's not like you run into this just anywhere."

"Connie, that's my point," says Brian. "That's my point. We are part of the emergence of this synthesis."

"It would happen to people our age," says Loyal.

"Definitely," Mary Evelyn concurs. "Thomas Berry is an exception."

Brian continues, "But now that the synthesis is starting to have some clear contours, it's quite possible that younger people will be activated and can enter into it without having to go through the whole long process of arriving at the synthesis. And that's the opportunity."

"The opportunity and the responsibility for us here is that it is not yet out there," I say, my zeal breaking through once again. "A couple of years ago I joined a Unitarian church in New York City, and soon after, I asked for the list of curriculum topics the central administration in Boston has as options for the youth program. I was curious about whether this, of all churches, would attend to the beauty of the story told by science and see that understanding as appropriate—if not central—for its youth. Under the environment heading, I saw several Native American lesson plans. But nowhere, under any heading, was there a single lesson plan that had any connection to the evolutionary epic—or to science, for that matter."

"Oh, I'd love to spend six months in your church and just preach that stuff!" Loyal practically jumps out of his seat.

"The problem is," I suggest, "you can't easily do it as an insider. You need to have someone who is regarded as an outside expert."

Mary Evelyn asks, "Well, you do something in your church, don't you Brian?"

"Yes, the same thing."

"What kind of church?" I ask, surprised to hear that this most eloquent activator of the evolutionary epic could find satisfaction in a regular church.

"It's a Catholic church," Brian responds.

Loyal raises his voice. "You're still a Catholic. I'm still a Lutheran. Now what the hell are we doing?"

"And I'm still a Presbyterian!" Ursula announces.

Loyal persists, "What the hell are we doing?"

"Ah," croons John. "Ritual is very important."

"It's cultural," Ursula insists.

"It's personal story," adds Brian.

"This is what we were talking about earlier," Ursula concludes. "You can't mess with that at all."

"If there were a Unitarian church in my little town. . . ." Loyal begins.

"Well, there is a Unitarian church in my town." says Ursula. "I went to it, but their ritual just didn't do it for me. So I went back to my Presbyterian church with my robes and my elaborate Christmas."

By this point I am incredulous. "I did not know this about you people! About Ursula's Presbyterian thing and Brian's still going to a Catholic church. What is that telling us? By the way, my Unitarian church is perhaps unusual because we have a big pagan component, and pagans do know how to stage a ritual. Our winter solstice celebration is fabulous—and that is, in fact, what drew me back to church."

Brian responds. "Well, I think in a certain sense it's telling us that with a hermeneutic that is large enough—this cosmic story—we can enter into relationships at all different levels."

Ursula jumps in. "Here's the end of the story about Jessica, now age 15. Although she doesn't get any personal therapy out of Christianity—we talk about that issue nonstop—it turns out we both are singers and we love to sing in the church choir. And so we participate in the service and rehearse every Thursday night for two hours with people for whom this is all the real deal. Like me, she understands that the ritual and the art and the music in particular in that tradition are terrific."

"I agree with you, Ursula, completely on singing," says Brian. "But all institutional life could and should be cosmological. If you think of cosmology here and institutional life over there, something's wrong."

Still reeling, I blurt out, "What I get from this confessional is that until the evolutionary epic has some institutions, some communities, some ritual, some *stuff* to offer people, it doesn't have a chance of being sufficient in itself. We don't just work with the old traditions in the meantime—indeed, if the accoutrements of this cosmology can ever come about in a self-sufficient way—we absolutely depend on those traditions, depend on them opening up in some way and allowing all this to enter."

The discussion continues, and then I shift the topic to one we had begun to talk about earlier but had left behind. "One thing I want to make sure we cover: I'm very interested in why we each think the evolutionary epic would be good for the world, not just for ourselves—as activators, Brian, not proselytizers. I'll start. Okay, at base I really care about biodiversity and the environment and so forth, and I also really care about culture and people's creativity, finding ways to be excited about the human project. So I worry that people are coming to regard our own species as purely bad news—that we're just a cancer on the planet. We environmentalists do have a tendency to promote that sense, though we try not to talk about it, and so we just do all the difficult, right things out of duty, some Puritan thing, not joy. Joy just isn't part of the program. And so for me, and what I hope other people, too, can pick up from the evolutionary epic, is the sense that it is through the human species that evolution has become conscious of itself, or as Jim Lovelock says, it is through us that Earth or Gaia becomes conscious of itself. And not just conscious. It's an aesthetic thing too. As Lovelock puts it, 'through the eyes of the astronauts Earth can now admire her own fair face.' Same thing with admiring all those galaxies Hubble now brings to us. It's not just, okay, we're conscious, okay, we got it—but, Wow!"

I continue, "So coming across you, Brian, and Thomas saying we're celebrants of the Universe Story—that's powerful! It's not just an intellectual thing that the sciences give us, but through our art and poetry and song we can now celebrate the story, the epic. Here the humanities join with the sciences in one big lovefest. That is what I would want people to pick up. That's why I want to tell the story, to be a catalyst. I'd love to inject that sense of excitement and celebration into our society, and I'm just optimistic enough to think that if that happens, something important will ripple around into the politics side as well as the personal therapy, and Earth will be better off because of it."

Mary Evelyn then ties in an earlier idea. "That's exactly why it's so important to work out, as I've been saying, the difference between imposition and evocation. We must evoke the story in others—not impose it upon them. This point is critical to the story taking root, in all kinds of ways we could never know, or expect, or even understand. To think that we could even begin to control the workings of this evocation is too much. But we can be catalysts. We must respect the process moving around us. As Thomas Berry reminds us, 'beware the paradox of conscious purpose' because conscious purpose can turn in on itself and blow apart. So our sense of being in touch with the unconscious processes of both ourselves and the Universe Story is crucial."

"I agree," says Brian. "I feel like my center of gravity is lower than my conscious purposes. I trust my sense of awe and joy most. I move with it and what evokes it."

"Brian, why do you do what you do?" I ask bluntly.

"You articulated it exactly just a minute ago," he responds. "Overall, it's something like wanting to participate in the evocation of well-being—the upper reaches of which are joy."

"To participate in the evocation of well-being," I repeat slowly, enamored of the words. "Not this top-down thing of 'I have the answer that is going to solve all problems!' "

"Now, it's also true that there are leaders." Brian continues, "All of us are leaders, and we have to recognize it. A certain responsibility comes along with that. But nevertheless, like Ursula brought up, it's more a feminist process, or a shamanic process— that's a word I like, too. It has that evocative sense."

The conversation moves into a discussion of strategies and niches—how we each see our own and the others' possible contributions, along with the sorts of talents or access this group is lacking. Among the things we agree we can do, however, is (as Brian has put it in an essay) "to call forth the poets." Others with artistic talents need to be sparked to tell the story. There is also a need for ritual. Brian speaks to this point. "One of the things Matt Fox says over and over again is that the vast majority of people don't get their cosmology from books. They get it from ritual, dance, theater."

"Earth Day is already there," suggests Loyal. "It's in churches now."

"I have a story," Brian continues. "Maia Aprahamian is a composer. I met her years ago, and then I heard from Thomas Berry that she was composing 'In the Beginning,' a choral piece based on *The Universe Story*. It was commissioned for the San Francisco celebration of the fiftieth anniversary of the United Nations. Anyway, she contacts me one day and hands me the score and asks me to be one of the narrators. There are maybe two hundred musicians and singers and dancers involved, with just a little bit of narration. I can hardly believe my own ears when I hear myself say, 'Yes, I'll do it.'

"So the thing happens. The place is packed. The music starts up and gets going. And I begin to feel—I really feel—what it was like to be in a primordial ritual. I get goose bumps, I even start to shake. The whole thing was just overwhelming. It really was. And I came away thinking that if we did more music, more art, more ritual, it would just completely ignite people."

Brian was surprised to hear himself accepting one of the narrator roles for the performance primarily because he does not feel adept at taking a leadership role in ritual. He loves participating in rituals, but he believes he has no talent for creating rituals or helping them unfold. Rather, "I'm a scientist focused on creating the epic, on telling the story, on writing the poetry of science." It is just great, however, if other people wish to ritualize the Universe Story, as with the U.N. celebration. But he himself engages in evocation strictly by writing and talking. For this conference, on Star Island, leadership or even participation in a closing ritual was out of the question for Brian simply because, by then, he would be on his way to another evocation (speaking) engagement. So when I approached him with a suggestion for a closing ritual, he encouraged me to pursue it and to look for help from others on the island. That's exactly what I did.

Celebrating the Epic

"You've got a great voice," I would say to a total stranger at an afternoon workshop. "How would you like to recite a little Robinson Jeffers or Joy Harjo for our closing ritual?" I had already lined up one of the IRAS board members, Bob Schaible, to recite some Walt Whitman. Bob, who teaches American literature, had led a week-long workshop on "Song of Myself" from Whitman's *Leaves of Grass*. I was not the only one who came away from his workshop convinced that Whitman was the bard of the evolutionary epic.

A closing ritual was not something to be performed by a select few. It should encourage participation by all. What sort of participation might this crazy mix of atheistic scientists, theologians, philosophers, members of the clergy, and strays like me find acceptable—perhaps even moving—and, above all, inoffensive? Billy Grassie, who teaches in Temple University's Intellectual Heritage program, offered to play the Pied Piper with his recorder, leading the group out of the dining hall after the banquet and into a semicircle on the lawn, facing the sunset and the sea. At the close of the ritual, he would lead the group once again, this time in a spiral dance—for this crowd, more a hand-in-hand walk than the whooping, racing spree that such a dance becomes at pagan ceremonies. We would draw from the pagan for this ritual, but it wouldn't *be* pagan. We would also draw from the Christian, but it wouldn't be Christian either. Hard-core science and American poetry would provide the substance.

"Instead of having everyone immerse an earth star in a separate cup, why not have the immersion take place in a communal basin?" Phil Hefner makes this suggestion when he overhears Billy and me discussing possibilities out on the porch.

"Great idea!" I say. "And how about if you, Phil, come up with remarks to lead the group into a silent meditation that precedes the earth star part? But no God or Jesus, okay?"

He agrees. Phil Hefner teaches at the Lutheran seminary in Chicago. He is an ordained minister and one of the father figures of IRAS, having served as president and now editor of IRAS's quarterly publication, *Zygon: Journal of Religion and Science.* His participation would ensure the wary that this event was meant for all.

"By the way," I ask Phil, "do you think anyone would be offended if I called this part of the ritual the Earth Star Communion?"

I had brought 150 earth stars with me to Star Island, figuring they might prove useful. I had encountered this remarkable organism first in the sand dunes of Oregon fifteen years ago, then in sandy glacial deposits of central Michigan, and finally all over our backyard in New Mexico. In dry weather these gray spheres are difficult to detect, but after a rain they darken and shape-shift. The stiff outer casing cracks and opens into a chocolate-brown star. Imagine a basketball splitting along the seams, but this one is only an inch wide and has a puffball mushroom at center. For years I have been delighting young and old with my magic trick of earth star immersion. Animal, mineral, or vege-

EARTH STARS, OPEN AND CLOSED. (Photo by Connie Barlow)

table? I ask. Most people haven't a clue. Learning the common name does not diminish the mystery.

So communal immersion it would be. In what? Star Island is very short of water, so we had all learned how to take sponge baths with a pitcher of cistern water and a plastic wash basin. Sacred wash basins, filled with sacred cistern water, would therefore surround our flower garden altar.

What about song? It would have to be something short and repetitive, easily picked up by ear as we move in toward the altar and then out again in the spiral dance. Judging from my own difficulties with hymnals, an ideal tune would span less than an octave. Here the pagans excel. I had learned a great chant-tune the previous winter solstice at my church, and the next day I redid the lyrics to tell the story of the supernova whose explosion created all the elements heavier than helium that make up Earth and our own bodies. Calling the supernova by the Babylonian mythological name Tiamat, suggested by Brian Swimme and Thomas Berry in *The Universe Story,* I had made this song the central feature of a ritual I designed for members of my family, with key roles played by two young nieces. For several years I had joined Dana and Eliza (with their father) in a candlelight Christmas Eve service and had sensed their longing when children their age starred in the manger scene and narrated the story. Christmas Eve was the only time these two went to church, as their father is a lapsed Catholic and their mother a secular Jew. Well, I decided, we can do our own ritual during the Christmas holiday—a ritual that tells the story of the Great Star Tiamat and thus how all the planets and the creatures of Earth came into being. My nieces loved the opportunity to make costumes and play-act and light candles, surrounded by zany aunts and uncles. The solemnity of the youngsters was astonishing. Because the adults enjoyed the Tiamat Song, too, I figured it might work for this Star Island group. We could introduce the song with a reading from the Swimme and Berry book and with a dramatic interpretation by Philomen Sturges, a children's book author and free spirit of the universe, who (as mentioned earlier) has renamed the Big Bang the Great Radiance.

Philomen was a hit. With a makeshift costume and a garbage-bag hat cracking in the wind, Philomen took us in two minutes from the explosion of Tiamat to the creation of Earth and on to the arrival of children in the universe. At this point, eight children came forward, teaching the Tiamat Song to the adults as the Spiral Galaxy dance began.

> Gift of Tiamat, out of the stardust we are born.
> Gift of Tiamat, out of the stardust we are born.
> Carbon, hydrogen, oxygen, nitrogen, sulfur, phosphorus,
> and trace elements.[39]

"That was fabulous!" Loyal Rue told me afterward. "I heard a couple of academics grousing a bit at the beginning of the dance, but by the end they were singing with gusto—especially the final phrase."

"The humor was so important," Mary Evelyn Tucker told me the next morning while boarding the ferry. "Humor is absolutely essential at this stage."

The ferry pulled away from the dock. The glasslike water, stilled by dense fog, slithered by while I revisited highlights of the week: the mind-to-mind magic that emerged from a six-way conversation, the morning chapel talks by Brian Swimme that I could enjoy while sitting outside on the rocks, the glorious hymns that spilled out of the chapel around ten o'clock each night, rediscovering Walt Whitman, my own workshop revealing the green passion of so many participants, the pulse of the distant foghorn, the sounds of the seagulls everywhere, the tingle of seawater when I joined the Polar Bear Club for an early morning dip, new friends, deeper connections with old friends. But mostly, I reran the ritual in my mind. My melancholy upon the end of a life experience was balanced by the exhilaration I felt for having made a real contribution.

3

Biology and the Celebration of Diversity

The evolutionary epic is the grand creation story for those of us who follow the way of science. The diversity of life, past and present, contributes all the characters for the Earth episodes of this epic. Plants, animals, fungi, protoctists, and bacteria alive today are the current players in a multi-billion-year, continuing saga. This pageant of life embeds our species in something far more magnificent than the comings and goings of cultures and kings.

The diversity of life here today is our extended family, and the very epic that reveals our wide kinship can also help us re-story those beings. In a sense, the diversity of life carries the outline of the narrative of evolution in a planet-wide gene pool, but only humans can be conscious of that story. Only humans, who have assembled and interpreted the fossil remains of trilobites and ammonites, of horsetail and seed fern forests, of dinosaurs and dire wolves, can celebrate the lines that have vanished—eons of particular stories exhumed from the grave. We can thus know and honor those who came before.

Eerily, too, we can look around today and know who among the finned and feathered and foliaged are truly the elders. For example, encountering a flush of avocado-green *Equisetum*, I am apt to conjure a foot-long Carboniferous dragonfly perched on a stem. A forest of tall conifers calls up a memory of their long-gone partners in evolution: the huge, snake-necked sauropods, such as *Diplodocus* and *Seismosaurus*. Some lineages of elders have changed so little in tens or even hundreds of millions of years that we call them living fossils. Living fossils receiving the most attention are those that now represent a single living twig on a long branch of the Tree of Life that has lost most or all of its other twigs. The hingeless brachiopod genus *Lingula* is such an example. This clamlike creature is almost indistinguishable from ancestral fossils of the same genus four hundred million years old. Brachiopods, now rare, were once as abundant as clams and mussels and oysters are today. Another living fossil is *Limulus*, the horseshoe crab, which spawns in vast numbers on beaches of the middle Atlantic states. The tanklike *Limulus*, with bulging compound eyes, bears a strong resemblance to those striking creatures of the Cambrian, the trilobites, which vanished at the end of the Paleozoic era. *Latimeria*, the coelacanth fish, was thought to have gone extinct sixty million years ago, as that is the age of its last fossil occurrence. But in 1938 it turned up in a fisherman's net, off the coast of southern Africa. *Latimeria* is related to the lungfishes, which, in turn, are almost identical with lungfishes recorded in rocks more than a hundred million years old. These lobe-finned fishes, moreover, provide a living model for the kind of fish that might, in its landward ventures, have given rise to amphibians and thence to all the terrestrial vertebrates, from reptiles to dinosaurs and mammals. Australia and surrounding islands are home to the only surviving lineages of an early experiment in mammal physiology, the monotremes; both the duck-billed platypus and the spiny echidna suckle their young by way of milk-producing mammary glands, but those young are hatched from eggs.

These elders, the living fossils, offer opportunities for connecting a celebration of biodiversity with the evolutionary epic itself. To re-story the diversity of life by way of science we thus can begin with the Old Ones.

RE-STORYING BIODIVERSITY

Look! There goes an Old One. Possum. Possum gives birth to fetuses and suckles them in a pouch on her belly. She has no close kin in this land. Almost all of Possum's relatives live in Australia. But once upon a time, Australia and South America and Africa and even Antarctica were all joined in one great supercontinent: Gondwanaland. Throughout that vast landscape wandered Possum's ancestors. They crept out of the trees each night to hunt insects and worms while the dinosaurs slept. Notice how Possum still stops and listens for the footfall of a great beast.

Look! There stands an Old One. Ginkgo. Ginkgo has no close kin at all, the last of a long line. Many millions of years ago Ginkgo lived everywhere. But then the climate changed and other trees came, and Ginkgo found refuge only on the other side of the world and at last in just a few mountain valleys and temple gardens of China. Today we honor Ginkgo in cities everywhere because Ginkgo remembers how to breathe air heavy with carbon dioxide. Look carefully at the strange pattern of branches and twigs, and consider how much effort it took trees to learn the best ways to accept the gift of the Sun.

Look! There clings an Old One. Lichen. Lichen pioneered the land way of life. Lichen turned rock into soil, then ceded the landscape to the stems and roots of plants. Today Lichen lives only where nobody else will—as spots and rings

GINKGO TREE. This wintertime photo reveals the strange branching pattern. (Photo by Deborah Winiarski)

on bare rock, as a crust coating the driest deserts, in patches on cold mountain peaks, on the rough trunks of trees. Contemplate the strength that such delicacy brings forth. Notice what comes of great patience.

The possibilities for re-storying biodiversity are as vast as the number of species that live on this planet. Thanks to the efforts of scientists and naturalists, it is possible to learn about kin on the other side of the world. A million and a half species have been formally named, and tens of millions more have yet to be given even this rudimentary level of attention. More than 4,600 species of mammals are known to exist, and new discoveries are still being made. In the past ten years the list of mammals known to science has grown by ten new species of primates, a new deer and wild ox, several bats, a new genus of antelope, and several genera and species of rodents.[1]

Science reveals the wonders of life in places no human has ever set foot (the abyssal depths of the oceans) and at scales too small for us to see (microbes). Science also alerts us to overlooked species close to home. Elders of intact indigenous cultures may well have names for all the bird species of their homelands, but, as Ed Wilson points out, they may make no distinction among ants.[2] An ant is simply an ant.

Through behavioral and physiological biology, we can discover astonishing facts about organisms that enhance our awe and reverence. Take ants, for example. As it turns out, the genus *Atta* became farmers (tending fungal gardens nurtured on minced leaves) long before we did. Careful observers in many parts of the world are surely acquainted with the stunning aerial skills of members of the falcon family and will have intuited that these birds of prey have keen eyesight. With the aid of science, we can also marvel at the ability of some falcons to track prey by way of the ultraviolet signature of rodent urine. Bees, too, can see in the ultraviolet, so some bee-pollinated flowers paint themselves with patterns invisible to our own eyes. Overall, science helps us sense what it would be like to *be* a falcon or a honeybee. How does the world present itself through alien eyes? How is the world apprehended through sensory organs we utterly lack—the antennae of a moth, the pressure-sensitive lateral line of a fish, the magnetic compass of a migratory bird? As Brian Swimme portrays the tragedy of extinction, the loss of a species is a loss to the universe of a particular way of *perceiving* a part of the cosmos.[3] Gone with the ivory-billed woodpecker is the ivory-billed woodpecker's way of perceiving the Louisiana swamp forest, the buttressed cypress in that forest, the fat grub beneath the bark, the beauty of an ivory-billed mate.

The ivory-billed woodpecker and many other species that were alive when I was born are now ghosts. It is the international reach of science that reveals the severity of today's extinction crisis. Science also puts this tragedy in perspective by telling us the full story of cataclysms on this planet. From this story we learn that our own assault on biodiversity is Earth's sixth major mass extinction. It has been sixty-five million years since the tree of life has been so violently pruned. Moreover, science reveals that this mass extinction differs from the previous five in shocking ways. First, a single species (us) is the root cause of the sixth great extinction. Feral goats and stowaway rats may be wreaking havoc on oceanic islands, but it is we who put them there. Pollution and dams may be the material causes of losses in aquatic systems, but it is we who created the wastes and impounded the waters.

Another chilling realization about this sixth great extinction is that it may well be the first mass extinction triggered by life itself, rather than, say, geological or other physical convulsions. Less certain, but a strong possibility, is that the sixth is the only mass extinction Earth has brought upon itself; possibly all previous biotic catastrophes are attributable to collision with a comet or a massive meteor.

Finally, distinctive in the present extinction is that plants are taking a beating. In each of the previous five, Kingdom Plantae suffered far fewer losses than did Kingdom Animalia. The likely reason for the survival of many plant lineages during previous extinction events is that plants can cling to life through rough times by taking refuge in seed or spore. Small reptiles and mammals can shelter in burrows and perhaps slow their metabolisms to the point of dormancy for several months. But there is no safe haven for giants. No terrestrial animal bigger than a turkey, in fact, is thought to have survived the end-Cretaceous cataclysm that sent the nonavian dinosaurs into oblivion. But plants today have little defense against an outright loss of habitat, owing to our own unquenchable desire to remake forests and grasslands into farms and lawns, to fashion marshes into parking lots, and to pronounce the rest pasture for our four-legged domesticates.

The distinctive flora of southern Africa is widely regarded as the world's most precious and vulnerable expression of plant evolution. Nearly three-fourths of the 8,500 plant species growing in the Cape's rich "fynbos" ecosystem are found nowhere else in the world. They are endemic to the region. Only a few tens of species of fynbos endemics are gone, but 600 are imperiled.[4] Perhaps the most insidious menace for flowering plants throughout the world is a loss of pollinators—bats and birds as well as bees and butterflies and moths.[5] Example: a Hawaiian bird with a long, curved bill succumbed to avian malaria unintentionally introduced when a flask of water containing mosquito larvae from abroad was poured into a Hawaiian stream. That bird extinction has consequences for an altogether different branch of life. Plants that had coevolved curving, tubular flowers joined the growing ranks of the living dead. Individual plants might live to old age, but they can no longer reproduce. Overall, if we cherish a flower, we must also cherish its pollinator and the predators that keep those who would consume it in check. If biodiversity is to be treasured, then so too is the web of relationships. Like Indra's net of gems in the Hindu myth, one can see reflected in even a small seed a universe of pollinator, predator, parasite.

Peter Raven has estimated that for every plant that vanishes, ten to thirty other organisms go down with it.[6] Consider not just the pollinator, for the moment, but the larval stage of an insect or two that may feed exclusively on that plant. A butterfly may have few loyalties as it flits from flower to flower in search of nectar, but it may have metamorphosed from a finicky caterpillar dependent on the leaves or buds of a single species. And that caterpillar, in turn, might be host for a parasitic wasp that eschews all other caterpillars. The wasp may be home to a parasitic worm that can find sustenance nowhere else, and the worm to an amoeba. At least two species of lice are known to have vanished with the last passenger pigeon.[7]

Some of us may be tempted to withhold our reverence for biodiversity from the ranks of parasites. But consider: Parasites foster biodiversity among the taxa we love. They ensure that no single species becomes too plentiful, thus providing room for others to find a similar niche. A disease-infested swamp may isolate antelope into two distinct populations that diverge through time, eventually giving rise to separate species no longer able to interbreed. Parasites may protect sibling species from hybridizing, thus ensuring that the two do not collapse into one. For example, hybrids of two kinds of cottonwood trees along the river where I live seem to be more susceptible to insect damage and hence less fit. Finally, some evolutionary theorists think that parasites are what drove life to invent sex—the scrambling of genomes at every turn is one way to befuddle the parasite's otherwise extraordinary ability to parry our every punch. Thanks to parasites and more congenial symbionts, biodiversity is far richer than it seems on the surface. In an essay titled "Endangered Interrelationships," Donald Windsor asks his readers, "When you are walking through a field and a deer pops up, how many species do you see? I see several dozen, from all the bacteria and protozoa in its gut, to the ticks, mites, and flies on its integument, to the fungi on or in its hooves."[8]

The sixth great extinction is upon us. How many expressions of life have we already exterminated? The answer depends on how far back one begins the calculation. Here is a list of prominent North American vertebrates whose extinctions are attributable to European occupation. Those for whom there are no more births include the Atlantic gray whale (extinct in 1750), Stellar's sea cow (1768), great auk (1844), giant deer mouse (1870), Labrador duck (1878), sea mink (1890), passenger pigeon (1914), Carolina parakeet (1914), heath hen (1932), Texas gray wolf (1942), and Caribbean monk seal (1960).[9] Since the 1600s, 484 animal and 654 plant extinctions have been documented worldwide. Many more species and subspecies—particularly the petite and retiring—have surely been lost without ever having been "discovered," and for many known species suspected of extinction, no one has yet put the time or money into documenting their absence. Contemplating the losses is the least upsetting way to bring the extinction crisis into one's circuit of concern. Far more depressing is to

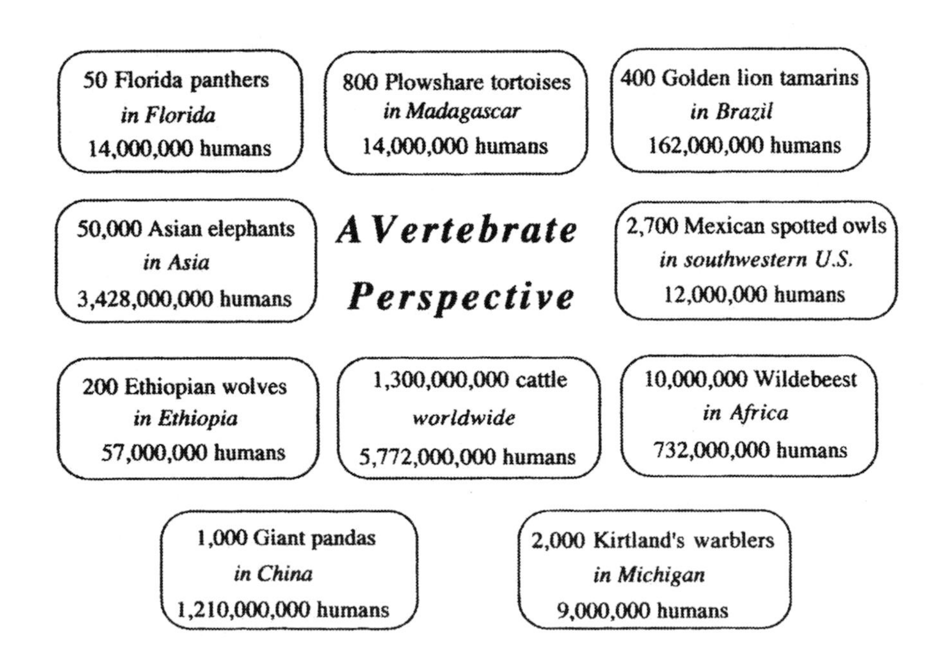

turn to those perched on the brink. A 1995 report of the United Nations estimates that an additional 5,300 animal and 26,000 plant species stand "a significant risk of extinction in the foreseeable future."[10]

These statistics constitute just the leading edge of the sixth major mass extinction, which actually began about 30,000 years ago in Australia. Aboriginal memory (confirmed by fossil evidence) includes some fabulous marsupial beasts that are alive no longer. Among them is *Diprotodon,* a rhinoceros-size herbivore that vanished about 20,000 years ago—well after humans first ventured onto the island continent. Beginning about two hundred years ago, with the invasion of Europeans, another wave of mass extinction ravaged Australia. More than half of the marsupial species that then inhabited the arid grasslands and billabongs of central Australia have since disappeared.

In the Americas, a wave of extinction coincided with the time when humans first colonized the Western Hemisphere. North America suddenly lost two-thirds of its genera of large mammals around 11,000 years ago; South America lost slightly more at the same time. Gone forever are the mammoths and mastodons, the giant ground sloths and tanklike glyptodonts. Gone forever, too, are the dire wolves and sabertooth cats, the short-faced bear and the largest species of bison, the American cheetah and the four-horned antelope, a huge hyena and a beaver as big as a bear. Gone (except in Africa) is the Golden Age of Mammals.

Distressingly, there is good evidence that we humans are to blame. Old World emigrants hunting their way through the Americas 12,000 years ago may have done a lot of damage to wildlife populations unaccustomed to bipeds armed with lethal projectiles. These earliest Americans might have killed too many of the large herbivores, the dwindling numbers of which then doomed the large carnivores and scavengers. The key piece of evidence supporting this Overkill Hypothesis[11] for the Americas is that Australia's marsupial mammals experienced a mass extinction 20,000 years earlier than the one that stalked the Western Hemisphere—and this was right around the time the first humans fanned out across that island continent. Consider, too, that ice ages had come and gone at least four or five times with no huge disruption of these same fauna, so how could the waning of the last glacial period have triggered such an extinction?

Biologist Jared Diamond and others link the end-Pleistocene extinctions, which took place on all continents except Africa, with the more recent extinctions that tracked the navigations of Polynesians throughout the South Pacific, ending just a thousand or so years ago. Half of Hawaii's native birds—including large flightless geese, flightless ibises, many species of flightless rails, owls, a hawk, an eagle, a petrel, and many small songbirds—went extinct before Captain Cook made landfall. Polynesians arriving on the shores of these previously uninhabited islands seem to have been responsible for the losses. Because oceanic islands are prodigious generators of new species, and because the Polynesians were such skilled seafarers, perhaps a fifth of the world's entire complement of bird species met their doom at least a thousand years before Europeans set sail.[12] Similarly, extinctions of large mammals, birds, and reptiles marked the coming of the Maori people to New Zealand and the first sailors (of Indonesian descent) to arrive in Madagascar.[13]

Diamond suspects that extinctions were inevitable whenever *Homo sapiens* colonized an unfamiliar environment (as in New Zealand and Madagascar) or advanced along a frontier (as in the blitzkrieg of the Americas). Extinctions also generally occur whenever people suddenly acquire a new technology (such as the guns of modern New Guinea, which devastated bird populations) and when wealth is concentrated in the hands of rulers of centralized states, who are out of touch with their environment. Finally, some species are simply more vulnerable to human acquaintance—such as the large, flightless birds of New Zealand and Madagascar.

We humans, whatever our ethnicity, are all culpable for these Pleistocene and recent losses. Even the greenest among us, if we had been born instead to those ancient times, would surely have participated in the unknowing slaughter of the most conspicuous and unwary mammals, birds, and reptiles. Jared Diamond judges that "tragic failures become moral sins only if one should have known better from the outset."[14] Our ancestors couldn't have grasped the cumulative consequences of their actions a thousand or more years ago; they were simply abiding natural impulses, enormously aided by culture-born technologies. Today, however, those of us with a scientifically rendered knowledge of historic and prehistoric extinctions are in a position to understand the causes and likely consequences of our own modern phase of biological plunder. The verdict this time around is therefore not manslaughter but murder.

The spiritual fare offered by science in re-storying biodiversity past and present encourages more angst than if we were unaware of the sixth great extinction. Only by way of science do we know that today's extinction tragedy is worldwide. Only by way of science do we know that our ancestors once walked with (and possibly stalked to extinction) giant ground sloths and elephant birds. On the other hand, science offers the condolence that there really was no "golden age of environmentalism."[15] We moderns are guilty, but we are not uniquely guilty. Throughout prehistory, human invaders worked horrors upon a landscape before they worked out ways to live in harmony, native to place. We who follow the way of science therefore cannot sink into an easy nostalgia. We know the only green way is forward. The golden age of environmentalism lies ahead—but only if we make it so. This is the story we must tell our youth. As Edward O. Wilson urges, "There can be no purpose more enspiriting than to begin the age of restoration, reweaving the wondrous diversity of life that still surrounds us."[16]

Islands Everywhere

The latest pre-modern wave of extinctions took place on islands: the myriad islands of the South Pacific and Hawaii (all colonized by Polynesians), the great islands of New Zealand and Madagascar, and many others. Islands are unusually susceptible to extinction, for while islands may generate a lot of new species by virtue of their isolation, those species tend to be fragile and their populations small. Island species have not been honed and tested by encounters with the robust species that the large continents breed. Humans are an exceedingly robust continental species. And the other species we bring with us on our boats—by design (as with pigs) or by accident (as with rats)—are robust, too. So-called exotic species (including us) transplanted to oceanic islands are thus the prime agents of biological holocaust.

Oceanic islands distant from land are fabulous evolutionary grounds for birds and reptiles. Gifted with flight, birds always find their way to remote islands. But once there, freedom from predators offers them an opportunity to shed flight muscles in favor of adaptations more suited to life on the ground. No longer needing to be airborne, some island birds become giants. The now-extinct dodo of the island of Mauritius is the most famous example; the dodo was a pigeon of uncommon proportions. The Komodo dragon is the most celebrated example of reptilian gigantism. Fortunately, this scaled-up version of a monitor lizard still thrives on several small islands in Indonesia, where it fills the niche of large meat eater in the absence of mammalian carnivores. This, the world's largest lizard, escaped extinction because of conservation efforts and, perhaps more important, because its hide is useless for leather making.

Mammals, with their high metabolisms and hence hefty requirements for food and water, are less suited than reptiles to survive a month-long journey on a log to a distant shore. But hippopotamuses, deer, and even elephants are capable swimmers over stretches of several miles. When they do take up residence on an island near a continent, evolution tends to make dwarfs of them. For example, pygmy hippos and elephants lived on the islands of the Mediterranean Sea until humans arrived and drove them to extinction. A dwarf elephant, now extinct, may once have been the Komodo dragon's prey. The islands off the southern tip of Florida are still home to the three-foot-tall, imperiled Key deer. The island archipelago of southeastern Alaska boasts a strong population of a diminutive deer. I recall once seeing a Sitka black-tailed deer swimming between islands when I lived in that region.

During the 1960s, Robert H. MacArthur and Edward O. Wilson pioneered a mathematical approach to understanding biodiversity in oceanic and near-shore islands. They built equations for calculating how the size of an island and its distance from the mainland ought to broadly determine the number of resident species. Smaller islands should have fewer species than larger islands, not only because of diminished acreage for keeping an immigrant population alive but also because less shoreline is available to intercept raft-riding or wind-blown adventurers. Likewise, the more remote an island, the less likely that an organism setting forth from the nearest continent will make landfall upon it. Less likely, too, is that the immigrant will arrive in good enough shape to begin the enterprise of making a living in a foreign land and that it will encounter a potential mate during its lifetime.

These causal explanations underlie the mathematical theory of island biogeography. Oddly enough, with the exception of exceedingly young islands, time is not an important factor in determining species richness. This is because biodiversity is not simply a stepwise accumulation of species over time. Even as one species is establishing a new life on a foreign shore, another species that has long since settled in, or even one that evolved on the island, is going extinct for one reason or another. An equilibrium number of species on a particular island should thus be evident through time. The species composition will change, but the overall richness remains roughly the same.[17]

The theory of island biogeography has tested reasonably well on real-world islands. Island biogeography has thus made a substantial contribution to theoretical ecology. But its real importance lies in what it can do for conservation biologists working with imperiled species on all the continents. Island theory can be used to determine how much habitat needs protection in order to ensure, over the long term, a viable population of an endangered or threatened species. A species need not, therefore, live on a bona fide island in order for island theory to prove useful. Forest fragments surrounded by a sea of clearcuts, patches of native grassland rimmed by cornfields, a stretch of free-flowing river constricted by a dam at either end—all these are islands, too. Such islands of our own making constitute the now-global plague of habitat fragmentation.

For part of the year I live among the forty "sky islands" of southern New Mexico and Arizona. Surrounded by impassable desert, populations of many forest- and stream-dwelling animals that retreated to these high elevations at the end of the last ice age are now confined to one island or another, without much, if any, intermixing. Sixty species are found nowhere else. A total of 175 endemics and nonendemics are considered imperiled on these sky islands.[18]

Elsewhere in the United States, forest fragments are more often artificial traps than natural refuges for species. Flying from New York City to Houston this year, I grew heartsick watching the scene that passed below for a full three hours. The great eastern forest was in pieces. Summed, the woodlots offer a lot of forest acreage for migratory songbirds that prefer a closed canopy. But small woodlots may be worse than no woodlots at all. A wood warbler might find the center satisfactory for a nest site and food and thus be tempted to end its northward journey there, but its eggs and nestlings become vulnerable to the small predators, house cats, and cowbirds (nest parasites) that thrive at the ecotone—the boundary between forest and field.

To better understand the effects of forest fragmentation on North American songbirds, the Cornell Laboratory of Ornithology put out a call in 1993 for amateur birders and other naturalists to participate in Project Tanager. I was among hundreds who volunteered to study any of four species of tanagers that breed in the United States and Canada. Both the hepatic and the western tanager live in my area along the upper Gila River in southwestern New Mexico. I chose to study the hepatic tanager because one of these brilliant red birds had trained me the previous summer to toss him bread on demand. This induced human behavior was especially crucial during the first few weeks of his parental duties, because finding food for hungry nestlings was difficult until the summer rains coaxed insects out of eggs and out of hiding.

In two consecutive summers, using a cassette tape of bird song that infuriated territorial males, I was able to confirm breeding hepatic tanagers in eight out of ten study sites. Unfortunately, so few volunteers lived within the hepatic's range (the American Southwest) that data for my chosen species were not statistically significant. Eastern birders did collect enough data on the scarlet tanager to determine that habitat fragmentation is indeed a significant problem for it.

The sixth great extinction is thus now spreading to the continents because we have carved up the continents into islands of habitat. A national park is an island of wildness surrounded by clearcuts, ranches, farms, housing developments, and roads. Only recently has it been noticed that national parks in the United States are generally too small to preserve their large (and sometimes even their mid-size) mammals over the long term.[19] Large carnivores, in particular, require huge and often exclusive home ranges. Habitat that can hold only a hundred grizzly bears, for example, presents dangers of inbreeding and the sporadic hazards of disease and weather that only a larger, more widespread population can endure.

In the Gila National Forest, where I live, the only hope for the endemic species of rainbow trout is fortification within habitat islands. Because the more competitive European brown trout is now entrenched in the main river, as is a species of non-native rainbow trout with which our endemic trout readily interbreeds, the Gila trout must be confined to upper tributaries with natural or artificial barriers preventing upstream movement of fish from the main river. For the fish's own good, therefore, we have relegated it to tiny fragments—aquatic islands—of habitat. Thus, although habitat fragmentation contributes to extinction, in some cases imposed fragmentation is a species' only hope.

One magnificent species ranging throughout much of the United States and southern Canada during the summer has put itself in danger of extinction by choosing just a few minuscule "island" habitats as its winter homes. This is the monarch butterfly. Except for a few monarchs that prefer Florida, the entire contingent of monarch butterflies breeding east of the Rocky Mountains winters in fir trees near the tops of just thirty mountains along a forty-mile stretch west of Mexico City.[20] The forested peaks officially are protected, but poverty and population pressures induce nearby residents to harvest timber and, formerly, to smoke the too-cold-to-fly butterflies out of the trees, the fallen bodies becoming food for cattle. Although numbering in the tens of millions, the monarch's evolutionary decision to put all its eggs in one basket, so to speak, puts it in grave peril. (The same holds for the western population of the monarch, which overwinters in just a few score dense aggregations—some a single tree—along the California coast.) In 1992, for example, a freak snowfall killed perhaps half the Mexican population of overwintering butterflies. The next June I read about the loss in one of my science magazines and phoned my sister in southern Michigan with the bad news.

"Oh," she cried, "that's why I haven't seen a monarch yet!" When some finally did arrive in early July, she made a special effort to gather twice as many eggs (about eighty) as she had been raising in years past. Each was given its own jar and a fresh clipping of milkweed leaf daily from a milkweed garden in her yard. The glass nurseries protected the caterpillars from the sometimes heavy toll taken in the wild by parasitic wasps. And when the butterflies emerged from their chrysalises, they were given back their freedom—to breed once more further north and east; their progeny would then make the long return journey to Mexico.

My sister is not a scientist; she is a homemaker who chooses to raise monarchs, which is something we did together as kids. The caterpillars advertise their toxicity with bold stripes and so are easy to spot. Because they eat only milkweed, and because milkweed is an early successional species (meaning it can be found along ditches and fallow farm fields), one doesn't have to grow up in a shining wilderness in order to make a friend of the monarch. Back around 1960, when Betsy and I raised our first caterpillars, all we knew about monarchs was that they showed up every summer and were fond of milkweed. We had no idea that our treasured neighborhood weedlot in Detroit was just one station along a magnificent journey that began and ended in central Mexico every year. Scientists didn't know that either until the mid 1970s, when the first Mexican enclave was "discovered" (local people, of course, were well aware of the butterfly wintering trees). Now, however, I am tuned in to the full monarch phenomenon, so spotting a butterfly on its way is one of life's great thrills. I encounter them regularly only on their southward migratory journey. Depending on whether I happen to be in the trailer we rent in New Mexico or the apartment we rent in New York, I will spot them either along the Gila River, strung out between the end of September and mid October, or streaming through Manhattan around September 23.

I remember one brilliant afternoon in late September, a few weeks after I had first moved to Manhattan. I was still struggling to adjust to an intense urban lifestyle, when, lo, out my seventeenth-floor window I saw an old friend flitting by, not two feet away. Here was a canyon maze of utterly blossomless towers, and yet this fragile being was making her way through, taking advantage of the thermals, perhaps, that rose from the southwesterly facing wall. And then came another, and another. I watched in awe and then rapture. For a moment, I loved those butterflies more than anything in the world. "Yes, yes, you can do it!" I cheered each of them on. I knew, by way of science, that these butterflies may have hatched in Maine or Nova Scotia, but evolution had directed them to head south in autumn. Unfortunately, evolution had not prepared the monarch for the obstacles of interstates across a flat and otherwise flower-rich landscape.

Two years later I was driving along the Iowa stretch of Interstate 80, slicing through a northward wave of returning butterflies that spanned more than a hundred miles. The parents of these midwestern wanderers had all overwintered in central Mexico, passing the baton to this next generation somewhere in between. I slowed down, grimacing each time I saw one enter the action. I counted the corpses along the road. And so I yearned for a bumper sticker that might change the world: I BRAKE FOR BUTTERFLIES. At a gas station beyond the killing field, I found a butterfly in my grill. Despite best efforts, I had surely killed at least three. I-80 was not a spiritual experience in a feel-good sense, but my anguish over little lives being lost welled up from the memory of that orange apparition of a Manhattan afternoon—and the magic of emergence I had witnessed in a jar some thirty years before.

Anguish, horror, guilt: These are the emotions all too often called up when we think about biodiversity today and the sixth major mass extinction. But there can be celebration, too. The city of Santa Cruz, California, knows how to celebrate. On Monarch Day in October, the mayor officially welcomes the butterflies to Santa Cruz, where a segment of the western population hunkers down for the winter in a few favored trees. The festival then begins, with music, poetry, and the by-air arrival of a costumed Monarch Man.[21] What if every town chose an organism and a special time of year to honor it? New York City surely could celebrate a Migrating Monarch Day in late September—and a Warbler Week in Central Park in May. What about slugs in Seattle, Maple Tree Day in Maine? Surely many regions could indulge in a hard-to-forecast but spontaneous day of Mayfly Madness. There are many opportunities for giving the human spirit a chance to connect with biodiversity and to balance numbing despair with a measure of joy and celebration.

A Conversation with an Earth Ecstatic

Diane Ackerman is a naturalist–writer who also has a special affinity with the noble monarch. "Monarch was the first butterfly I knew by name," she recalls in her 1995 book, *The Rarest of the Rare*. "Like most children, I found them magical and otherworldly, a piece of the sun tumbling across the grass . . . Fluttering madly—but moving slowly—from bloom to bloom, they looked the way my heart sometimes felt."[22]

Diane Ackerman not only wrote about the monarch; she worked to protect it. In 1986 she tagged butterflies at 65 overwintering sites along the California Coast and wrote about monarchs for *Life* magazine, hoping to persuade California lawmakers to protect the scattered groves of coastal trees that are winter homes for the entire population of monarchs breeding west of the Rockies. Protective legislation passed in 1987.

A decade later I talked with Ackerman. Her collection of essays on endangered species, *The Rarest of the Rare,* is a lush synthesis of science, adventure, and aesthetic reflection. Unlike many epistles on endangerment, however, the grim reality Ackerman enters into does not make her a dour guide. Helping scientists tag and measure monk seal pups in Hawaii or reintroduce captive-bred golden lion tamarins into the wild, her attention is wedded to the rich details of the sensuous present. "Wonder is the heaviest element on the periodic table," she reflects. "Even a tiny fleck of it stops time."

"I'm grateful that there are outspoken doom and gloom nature writers," she tells me, "but that's not what I choose to do. My strategy is to celebrate. I believe that if you can cause someone to fall in love with an animal or a landscape, they won't want to lose it. They will fight to protect it."

That's exactly my outlook, too. "There's already a lot being said about our moral duty to other species," I comment. "That's important. But without the celebration, the joy, bad news threatens to plunge us into denial or despair. So much more is possible if we become charged with a kind of religious conviction that biodiversity matters—that biodiversity is sacred."

"I think more and more people are searching for such a conviction," she says. "I believe there's a terrible hollowness, an emptiness at the core of society right now that comes from our trying to exile ourselves farther and farther from nature. Nature is something that most people visit on weekends. Yet we evolved to be intimately tied to nature, to feel whole and natural when we belong to nature, and to respond to the ever-changing fantasia of the seasons. The harder that we try to deny that heritage, the more uncomfortable and alienated we become."

The conversation circles around the act of writing, jotting down sensory impressions in the field, capturing a metaphor, and engaging in what Ackerman calls "the poet's quest." She reflects, "I think of myself as a nature writer if what we mean by nature is the full sum of creation, and that applies to all the genres in which I work. It doesn't matter if I'm recording an expedition to a distant land, describing the habits of squirrels, or writing about the life of a seventeenth century Mexican poet (who was also a scientist and a nun). It doesn't matter if I'm writing about the dark night of the soul [as she does in *A Slender Thread*, 1997]. My quest is always the same. It's a poet's quest.

"Poets are people who tend to be bothered by disturbing questions. There are two questions that bother me a lot. One of them is, How do you start with hydrogen and end up with us? That is, How do you get from the Big Bang to the whole shebang? The second question is, What was it like to have once been alive on the planet? How did it feel in one's senses, passions, and contemplations? Everything I've written, in fourteen books thus far—poetry and prose—has been an attempt to find answers to those two questions. Deep down, I know they should take from birth to death to answer and include all of consciousness, all of life."

"Something comes through in your writing that is a kind of fearlessness," I note.

"Liberal religious authors I encounter often show gratitude to science for providing not so much answers as more and more mystery. These authors revel in an enchantment with mystery and find much of their spirituality there. But at the same time I detect an underlying reluctance to be fully and completely open to everything that science may reveal. There's a worry that some answers science might produce could bring spiritual discomfort. In contrast, I get the sense that for you, there's nothing you could ever learn, nothing you might witness in nature, that would cause you to turn away. Somehow you would find a way to celebrate it."

"I think that's true with one exception," she responds. "And that is that I find it uncomfortable, heartbreaking, disarming, frightening, and appalling in a hundred different ways when I learn new forms of evil or cruelty that human beings exact on one another. Other than that, you're right. I am open to the truth of the Earth. I'm a great fan of the universe."

I now turn to the question that is foremost in my mind. "In *The Rarest of the Rare*," I begin, "a religious sensibility comes through in your encounters with endangered species, in a kind of communion with nature. Would you, in fact, describe yourself as religious?"

Ackerman responds without hesitation. "I've always felt an ecological spiritualism, and this powers everything that I write. When I was growing up, I was simply curious about all religions, all peoples—everything. As a teenager, I read about the lives of saints and explored Asian religions. These days I may read a book by a Jewish mystic, or by a Buddhist, or by other religious folk, as well as books by novelists and poets and nonfiction writers. I don't feel satisfied by any organized religion that I've encountered. Yet I'm probably the most religious person I've met. I'm deeply religious. It's just that I don't require a governing god in my sense of the sacred."

I am thrilled. This is fully the answer I had hoped for from one of my writing heroes. I explain, "One of my mentors, who unfortunately died before I was intellectually ready to encounter his ideas, wrote a book called *Religion Without Revelation*. This was the evolutionary biologist Julian Huxley, grandson of Darwin's great ally, Thomas Huxley. The religion Julian promoted was a celebration of the evolutionary epic. What struck me was that even though Julian was a nontheist, he wanted to claim the word *religion*. He didn't want it to become the exclusive property of theists. But when I interview scientists today, I find I have to be careful about using that term."

"I don't have any problem at all with religious language," she says. "I cherish the origins of religious terms. *Holy* we can trace all the way back to the Indo-European. It meant the healthy interrelatedness of all living things. From this we get our word *whole*. As a result, I have no trouble using a word like *holy* to describe a place in the wilderness where I might feel an intimate relationship with the cosmos."

"You use plenty of other such words, too," I add. "*Benediction*, *sacred*, even *prayer*—all these words come out in your essays. But the most striking religious term is the one you invented." I reach for my bookmarked copy of *Rarest*. "Here it is in your chapter on the gravely endangered short-tailed albatross: 'I go in part to stand witness. Life forms such as these need to be beheld and celebrated. That is my privilege as an Earth ecstatic, but it is also my duty as a member of the species responsible for their destruction.'

"*Earth ecstatic*: that's a fabulous identity to take on," I conclude.

"It's a personal religion that fulfills me in countless ways," she responds. "My creed is simple. I believe in the sanctity of life and the perfectibility of people. I believe we should regard all life forms with dignity, respect, affectionate curiosity, and the kind of protectiveness family members feel for one another."

Life Loves Life

Our conversation continues: "I noticed in your introductory section to *The Moon by Whale Light* you say that when asked whether you prefer whales or penguins, you respond, 'I prefer life.' I'm curious. I know I have shifting allegiances to different organisms, depending on where I happen to be and what I am doing. My allegiance right now is to hepatic tanagers, as I have been volunteering in a nationwide field survey of tanagers. So I'm wondering, when you're in the midst of working with scientists on some particular animal, do you find yourself in alliance with that creature, and then do you move on to the next?"

Ackerman responds, "I do fall in love with whatever animal I'm working with, but as I said, I prefer life. I have a poem called 'Ode to the Alien.'[23] It's an old-fashioned love ode; it just happens to be written to an extraterrestrial. But it's really an ode to all life forms that seem alien to us. I end it by saying that life loves life."

Life loves life. That sounds like biophilia—an idea promoted by Ed Wilson, who played a prominent role in the previous two chapters and who is also one of the originators of the theory of island biogeography. I have a hunch that many of us who deeply value biodiversity, who find it sacred, have fired that conviction with the memory of an intimate encounter with one or more real organisms at some point in our lives. These organisms are usually not the dogs and cats that have become honorary humans, but creatures that are still alien, that are at home in the wild and can therefore never be fully at home with us—try as we might.

I remember trying often as a child, and with many different creatures. A "rescued" baby cottontail would hide and pee behind the furniture. A tortoise would refuse to eat. A spotted salamander would find refuge in my grandmother's shoe, a snake disappear into a drawer. A cecropia moth would emerge too soon from its cocoon in the unseasonable warmth of my room and haunt my mother at the bathroom nightlight. A neglected crayfish would creep into the living room in search of a creek. I could not intuit or measure up to their needs. So back into the wild they would go.

Field scientists are the lucky adults authorized to maintain their childish ways. They can hold onto an intimate involvement with wild things—sometimes very intimate. When Diane Ackerman accompanies scientists into the field she goes as assistant, not observer. She wrestles a monk seal for measurement and tagging. To sex a crocodile, she slides a finger up the cloaca and then reports to the rest of us how it feels and smells. Yes, the work is objective and follows research protocols, but a thrill and a bonding are not thereby precluded. Here is a sampling of how some of the scientists you will meet (or have already met) in this book answer the question "What is your favorite organism?"

"My wife," responds Ed Wilson finally, after a long pause. "But I suppose you mean species or groups of species. In that case, well of course my answer would be ants." He continues, "I'll tell you a test you can apply to any naturalist. And that is, if they hesitate, as I just did, ask them what they would look for in the way of organisms if they were put down anywhere in the world. And my answer, absolutely immediately, no matter where you put me, I start looking for ants."

"Even in the airport," I add, remembering a story he told in his autobiography, *Naturalist*.[24]

"Yes. I'll put down in an airport and if there's a little bit of a garden there, right next to the tarmac, I'll immediately be poring over it, looking for them, figuring out what's there. Even sooner, when the plane touches down and begins to taxi, I'm looking out the window, searching for ant mounds."

We share a laugh. "Isn't it wonderful to be that way!" I say. "Can you imagine *not* being that way?"

"They're my *friends*!" Wilson exclaims. "I know them, intimately. Everywhere I go in the world I'll find ones that I know, my friends, and I'll find ones I haven't seen before."

"I tend to be drawn to small, secretive forest ruminants." This is Stephan Harding speaking, an ecologist who will play a role in the Gaia chapter. Harding continues, "Not the big ones—but the little ones, like the muntjac deer I did my Ph.D. work on or, even better, the mouse deer. The mouse deer is a beautiful, small—only about rabbit size—ungulate in the Old World tropics. In the neotropics I'm drawn to the ecologically equivalent caviomorph rodents."

"Do you know why you are drawn to these creatures?" I ask.

"It's very difficult," he responds, "but I think it has to do with the way they move through the forest. The fact that they're small is important. If you are looking at a big animal in an ecosystem, the scale is too big. They have huge home ranges, and so you can't quite find out what they're doing. You can't follow them. And if they're too small, like mice, they disappear under leaves and things. So you lose them that way. But for the mouse deer or for the muntjac, you can stay with them for long periods of time. And so you begin to see other parts of the ecosystem from their perspective. A single bush becomes a quite significant thing in their home range. Whereas a bush in the home range of a moose is perhaps not terribly important. But in the home range of a small ungulate, a bush is a juicy, succulent, significant item. And also the beauty of them, their delicacy, moving in and out amidst bushes. Because these small creatures are sometimes hard to spot, when they do appear, they suddenly transform the forest into a place of beauty."

Harding recalls a particular experience with the animal he studied while at Oxford. "The Chinese muntjac, which now also occurs in southeast England, was brought over by an aristocrat at the end of the last century. He let it loose on his estate. Now there are masses all over this part of the country, and they've actually become a pest. One day when I was a student, the porter [at one of the colleges in Oxford], who knew of my work, called me when a muntjac came onto the grounds and started munching roses. He said, 'Come over quick and catch this deer!' I caught it in the pillow my grandmother knitted for me."

"How do you feel about loving a weed species?" I ask.

"There's no such thing as weeds," he chastens me. "Even for plants, I never use that term."

"So when we bring them into places," I venture, "it's our moral problem, not theirs."

"Exactly, and we should respect them because they are just trying to live. To remove them from an area is fair enough, but we should do it with respect."

"My current favorite organism is *Calonympha*," responds Lynn Margulis. She is a biologist renowned for work on the role of symbiotic mergers in evolution and as a proponent of the crucial role played by microorganisms in Earth's physiology.

"*Calonympha* is a protoctist," she continues. "More than that, it's an archaeprotist, which means it belongs to the earliest evolved organisms with nuclei—those without mitochondria. But it has many nuclei, up to fifty or so, in its single-celled body. And it has four undulipodia for every nucleus. Since *Calonympha* is the 'big microbe' in which we may have found centriole-kinetosome DNA, of course I love it!"

Finding DNA in the centriole-kinetosome of this creature is thrilling for Margulis because it further corroborates her theory of symbiogenesis, in that yet another set of the structures within complex cells are less attributable to mutations of the nuclear DNA and more the result of symbiotic mergers with smaller bacteria that, like all cells, possessed their own DNA. The one or two billion years that have passed since the symbioses started obscure much of the evidence, but genetic remnants of the bacterial intruder may yet remain within the cell structures of descendants alive today.

"What did you love before you began to work with *Calonympha*?" I ask.

"I was then and am now very fond of the animals in which *Ca-*

lonympha and its relations live, the termites *Cryptotermes brevis* and *Cryptotermes cavifrons*. These wood-eating termites, at least *C. cavifrons*, harbor another protist whose entire head end revolves on its posterior. It is aptly called 'rubberneckia' by Sid Tamm, who beautifully described it. We can't italicize the name because this archaeprotist has not yet been formally introduced into the scientific literature. All sorts of other marvelous microscopic creatures abide with rubberneckia and *Calonympha* and probably have for over a hundred million years. So termites fascinate me. And they always take me to beautiful places. To find them I go south and into wet habitats."

"So what you really love is the little community in the hindgut of termites," I surmise.

"Yes." She continues, "But I've always loved plants, too. In general I like plants and fungi much better than animals."

"There's a beautiful little hepatic, a liverwort: *Diplothyllum apiculatum*. It was a stunning teacher for me when I first met it in graduate school, and it is still beautiful." This is Paul Mankiewicz speaking, a biologist who coaxes plants and microbes to restore broken landscapes in a region that many would find difficult to love: New York City.

"What struck you about this creature?" I ask.

"The way the leaves are folded has a certain kind of perfection to it, symmetries on asymmetry. It is stunning also in that it won't grow normally in culture. I could not get it to grow recognizably in sterile culture. Instead of an organized plant, there were heaps of cells."

"Well, what does it need?" I ask.

"If only we knew!" he responds. "But really, to your overall question about a single favorite organism. That's just off the map. Even so, I always have loved this one."

"So it's a stupid question," I say, "but you can nevertheless answer it for me."

"Kestrels are absolutely exquisite creatures," he continues his reverie.

"My god, they're vertebrates!" I joke, knowing Paul's penchant for microbes, fungi, and the tiniest plants—all of which play key roles in the cycling of nutrients in ecosystems. Kestrels, in contrast, are small, predatory falcons that live off the productivity of other living beings. To the human eye, however, these masters of flight (that can also see in the ultraviolet) are beautiful to behold.

"It's true; they are vertebrates. But holding that against them is a bad position. The fact that they're vertebrates obviously means that biochemically they're just variations on a theme. They can do virtually nothing. They are entirely dependent on a whole set of food chains, a whole set of microbial systems."

"But they're charming despite their uselessness?"

"All they can do is control other vertebrates and some invertebrates. It's a macrofauna game, basically—played by a minority of organisms on the planet, all larger than a centimeter or so. The control of other macrofauna does make it possible to have a greater mass of phosphorus available for everybody else. So by eating voles and grasshoppers, kestrels do change the structure of the ecological fabric. In general, what's good for the sparrow hawk is good for the plants."

Paul is a restoration ecologist. He tries to nudge damaged landscapes back into a state of health. His focus on the ecosystem services that living beings perform will become very evident in the next chapter. I ask him, "Well, is that lovely bryophyte of yours any more useful?"

"It increases the capacity of the entire microbial system by enlarging the structure and enhancing the interface of surface areas. It is excellent scaffolding for microbial attachment, but also think of its water-holding capacity. Before bryophytes evolved, there were no large-volume capillary spaces in the terrestrial biosphere. But now, with bryophytes and the far bigger newcomers, an immense volume of water can be stored and later used."

"I see your point," I concur. "Bryophytes were the organisms that gave to the microbes something that the microbes could not do for themselves."

"I definitely have a totem relationship to the Shasta ground sloth," responds Paul S. Martin.

"My favorite extinct organism!" I exclaim. "I dedicate my book to giant ground sloths!"

"I've got a model of one outside my door; it's a life-size wire model," he continues. Martin's door is at the Desert Laboratory of the University of Arizona in Tucson, where, after more than three decades on the faculty, Martin is a very active emeritus. His specialty is paleoecology, and he is the originator of the Overkill Hypothesis for the end-Pleistocene extinction of big mammals. You will hear more about his work and enthusiasms in the next chapter.

"How did you develop a totem relationship to the ground sloth?" I ask.

"Through hunting for sloth caves and collecting sloth shit," he explains. "When I first came to the Desert Lab in 1957 to work with Ted Smiley in the new program of geochronology, the very first day he showed me a paper bag of sloth shit. I thought to myself, 'I've come to the right place!' "

"So it was love at first sight when you saw your first dung ball?"

"Oh, yes!" he responds.

"I was actually asked that question—about finding a favorite organism—at a very important time: the first year of graduate school." Lee Klinger, an ecologist who will reappear in the Gaia chapter, tells his story. "In fact, I wasn't even asked the question; I was told that to be successful as a biologist, I would have to pick an organism. Not only pick an organism but become, literally, the expert on the organism—a single species. Every biologist does this, I was told. Most of my fellow students already had their favorites. So I started to think about it. 'Okay, what's my organism?' Well, I liked *Dryas* [a flower of alpine and arctic tundra], but as to my favorite? I could not decide. I just could not decide. It was at that moment that I realized I'm not a biologist. So, a year and a half later I ended the program with a master's degree and went on to a totally different department, Geography, for my Ph.D.

"To this day," Klinger continues, "not only can I not pick a species; I can't even pick an ecosystem. You would think it's bogs, but remember, I'm interested in the formation of bogs. So when I go into a bog, it's beautiful being there, but I don't see that much evidence in the bog itself of how it formed. If I want to know how a bog formed, I have to go into the forests that surround the bog and look for the processes, the transitional kinds of phenomena that lead to bogs. So I actually spend most of my time in the areas around peat bogs—not in them."

"How did you choose to get close enough to this ecosystem to develop a deep affinity for it? Why not tropical rainforests or deserts?" I ask.

"That's a story! I worked in Southeast Alaska as a logger for a couple of summers while an undergraduate in college. I had been involved in environmental organizations at school, yet I lived in a log cabin. I mean, I built my house from wood, so how could I be against logging? So I went out to be a logger for a while."

"Cutting down big trees in Alaska—how did this make you feel?"

"It wasn't the individual trees I had a problem with," Klinger responds. "It was the feeling of what was happening on the whole hillside. I could do anything to an individual tree; it wasn't a big deal. It's the landscape that really made me feel that there was a problem. But I learned an incredible amount. It's a catastrophe what's going on up there, like I had suspected. But I didn't know. I did not know. Actually, the forest industry had some valid points. They said, look, if you don't disturb the forest, it starts to deteriorate; it just turns to bog. And the environmentalists were ignoring this."

He continues, "What that time as a logger did for me, as well, was cement my love for this area: Southeast Alaska. Early on I decided that this is where I wanted to do my graduate fieldwork. I remember hiking through the old-growth forests and seeing this incredible biomass of moss. Everybody else was studying the trees, and I just said to myself, 'I know there's a story here.' I sensed it had something to do with the mosses, about the interaction of mosses and trees."

Exploring that interaction has become his life's work. Lee Klinger now studies forest–bog interactions all around the world—and not just in the colder climes. "The first time you saw a bog in the tropics," I asked him, "were you thrilled?"

"Oh God, Costa Rica! It was like I was back in Alaska. I was finding a lot of genera of mosses and lichens and even some vascular plants that were the same. That's just on the taxonomic scale; morphologically, there's even more similarities."

Scientists are not the only people with attachments to particular organisms or natural processes. Identification with creatures is deeply rooted. For example, an individual or a clan within an indigenous culture is likely to have a totem organism. A Lakota boy in America will be launched on a vision quest, hoping for a sign of bonding and a power relationship with an animal ally. An Aboriginal Australian will take on the "dreaming" of the ancestor associated with the geographic place at which his or her fetal self quickened and was first felt within the womb. These ancestors have an animal nature: Kangaroo Dreaming Man, Tortoise Woman, Emu Woman, Little Wallaby Man.

Strong associations with animals linger even in my own culture. Baseball teams include the Detroit Tigers, Baltimore Orioles, St. Louis Cardinals, Toronto Bluejays. Football has its Lions, Bears, Rams, Dolphins, and Seahawks. Even plants have a place: Maple Leafs is the name of Toronto's hockey team. In the United Kingdom the national rugby teams have these symbols on their uniforms: a rose for England, a thistle for Scotland, a leek for Wales, and a shamrock for Ireland.

Those of us not associated with a university or institute in which an e-mail address is created for us have an opportunity to take on an alter ego, which could very well be a favorite organism. Stephen Harding, for example, defies professional custom by combining his initials with the name of a small forest ruminant: agouti. The Internetted segment of my family is a good example. My sister goes by Metalfrogg; she thrives on heavy metal music and, having recently moved to the monarchless rainforests of the Cascade Mountains, now raises tree frogs as a hobby. Metalfrogg bears two g's because some other subscriber to her communication system had already taken on the name Metalfrog. I had a similar experience when I tried to christen myself Tanager at the time I went on-line. That name was already in use, so I became CBTanager. My sister's son has chosen Iggy, after his pet iguana. My brother's two daughters have chosen e-mail names that merge their given names with "horse" and "pup."

Some of us have bonded with creatures we have never met. I recall an Earth Day service at my Unitarian-Universalist church on the upper west side of New York City. Our minister led a guided meditation for those of us who wished to join him in finding or enhancing a spiritual relationship with another species. At the end of the meditation, he invited any who so wished to call out the name of their chosen creature. "Squirrels" was a popular choice for some in that very urban audience, but I was surprised to hear more than one dolphin and eagle call out, and three or four wolves.

That's the depth of identification needed to turn around this extinction crisis. And it was a church that promoted it. Another church in New York City known for its spiritually green ceremonies (on Saint Francis Day, Earth Day, and the solstices) is a grand Episcopalian church, the Cathedral of St. John the Divine. Each year thousands attend the Winter Solstice celebration in that awesome stone building, the program designed and hosted by the musician and composer Paul Winter. The highlight for many of us is an opportunity to add our own voices to a thundering chorus of howls, as Winter and his tenor sax transform into the leader of the pack on this darkest night of the year.

Joanna Macy and John Seed have taken this potential for human identification with other species several steps further than the howling spree I enjoyed in church. In their "Council of All Beings" workshop, participants are encouraged to think deeply about what it means to be a particular creature and what human-imposed dangers are threatening its well-being.[25] In his opening invocation, Seed offers,

> We call upon the spirit of evolution, the miraculous force that inspires rocks and dust to weave themselves into biology. You have stood by us for millions and billions of years—do not forsake us now. Empower us and awaken in us pure and dazzling creativity. You that can turn scales into feathers, seawater to blood, caterpillars to butterflies, metamorphose *our* species, awaken in us the powers that we need to survive the present crisis and evolve into more eons of our solar journey.
>
> Awaken in us a sense of who we truly are: tiny ephemeral blossoms on the Tree of Life. Make the purposes and destiny of that tree our own purpose and destiny.
>
> Fill each of us with love for our true Self, which includes all of the creatures and plants and landscapes of the world. Fill us with a powerful urge for the well-being and continual unfolding of *this* Self.

> May we speak in all human councils on behalf of the animals and plants and landscapes of the Earth.

May we speak in all human councils on behalf of the animals and plants and landscapes of the Earth. This is the heart of the Council of All Beings ritual, an unfolding and spontaneous creation. Identification becomes so deep that participants speak out not *for* an organism or landscape but *as* a fish, a fern, a mountain. More than a decade ago John Seed, an Australian activist, stunned many environmentalists by claiming that he was not actually working to protect Australia's nub of native rainforest. Rather, "I am part of the rainforest protecting myself."[26]

As conceived by Seed and Macy, scientific knowledge is an integral part of the meditative venture for expanding the local self into the greater Self of another creature and the world as a whole. "There is science now to construct the story of the journey we have made on this Earth, the story that connects us with all beings. Right now we need to remember that story—to harvest it and taste it. For we are in a hard time, a fearful time. And it is the knowledge of the bigger story that is going to carry us through."

It is by way of this story, they tell us, that one discovers "the powerful erotic energy of evolution." It is by way of this story that one's very being can expand and fill with a passionate caring for the vast diversity of life—past, present, and future.

Ecology and the Birth of Bioregionalism

"We could say a food brings a form into existence. Huckleberries and salmon call for bears, the clouds of plankton of the North Pacific call for salmon. The sperm whale is sucked into existence by the pulsing, fluctuating pastures of squid."[1] Gary Snyder thus provides rich images of biodiversity in context, which is a key aspect of ecological understanding.

Ecology is the science of relationships. It includes the relationships of a living being to the chemistry, climate, and topography of its home. Ecology is also relationships among the living: of herb and herbivore, prey and predator, flower and pollinator, host and parasite, and all partners in symbiotic affairs. A living relationship is not so much action and reaction, nor even interactions—more, transactions.

Ecology is thus the busy commerce of living beings. Paul Shepard extolled ecology as the science in which "relationships are as real as the thing."[2] Gregory Bateson maintained there was no such thing as an organism, only "organism-in-environment"[3]—no separation.

At the heart of ecology is the ecosystem, in which living and nonliving participants forge a greater whole. The boundaries of an ecosystem are to some extent in the eye of the beholder; an ecosystem can be expanded or contracted pursuant to one's perceptual aims. An ecosystem, in turn, becomes a bioregion—again, flexible to various scales of interest—when transposed into the realm of felt homeland. Bioregion is an ecosystem beloved.

Bioregionalism is the green equivalent of multiculturalism. The bioregional movement got its start in the 1970s through the efforts of Peter Berg, Paul Shepard, Gary Snyder, Stephanie Mills, and others.[4] These leaders advocated a knowledgeable and loving attachment to one's home region. Foods, building materials, water, energy, and other natural resources were to come from the bioregion in which they were to be used. The methods of harvesting resources and returning wastes were to be sustainable, ecologically responsible, and sensitive to the needs of other species inhabiting the bioregion. David Abram, after the American poet Robinson Jeffers, characterizes the bioregional movement as the act of "falling in love outward." Bioregionalists are those who "dedicate themselves to the terrain that has claimed them" and who resolve to "meet the generosity of the land with a kind of wild faithfulness."[5] I envision that someday each bioregion will have crafted its own unique final chapters of the evolutionary epic, relevant to *this* place, *these* creatures.

Broadly speaking, it is the science of ecology—physiological ecology, behavioral ecology, population ecology, community ecology, ecosystem ecology, landscape ecology—that provides the understanding essential for devising ways to live in place with minimal impact. The sciences can also enhance a sense of intimacy with place. Philosopher Baird Callicott observes that various sciences "supply substance to 'scenery.' "[6] The beauty of a river, a mountain, a desert is enhanced by an engaged and knowledgeable mind. Overall, bioregionalism is a scientifically informed movement for hunkering down. In a world in which problems often seem too big for anyone to solve, bioregionalism can be an attractive way to channel one's zeal for informed participation. At the bioregional scale, one human being can, in fact, make a difference. Bioregionalists can think locally, act locally, and expect to see some results. An individual can circumambulate the watershed; that same individual can make a difference in its protection or restoration.

A bioregion is defined not by political boundaries but by ecological coherence. It encompasses a suite of life, along with the rocks and waters, the lay of the land, and the climate patterns that shape the region. Watersheds are commonly called upon to delineate bioregions, as in the Upper Gila Watershed and the Hudson River Estuary. These are my two bioregions, more commonly known as southwestern New Mexico and New York City. (Why two home bioregions? Ask the hepatic tanager and other migratory birds.)

One of the most important actions bioregionalists can take is protecting the biological treasures that have more than regional significance. Science can reveal who among us is endangered and which bioregions bear responsibility for their survival. In the Upper Gila Watershed, which is blessed with vast public lands managed by the U.S. Forest Service, animals that have been designated "endangered" or "threatened" by federal or state agencies include two snails, one snake, four fishes (among them, Gila trout), a dozen birds (notably, Mexican spotted owl and southwestern willow flycatcher), and four mammals (spotted bat, Arizona montane vole, desert bighorn sheep, and shortneck snaggletooth). Merriam's elk is not on the list because this native subspecies of elk joined the ranks of the extinct earlier this century, a casualty of overhunting. Elk were reintroduced here fifty years ago, and they are now thriving, but this is the Roosevelt subspecies, which is native to Yellowstone National Park much farther north. Perhaps in a year the Mexican gray wolf will be added to the list of endangered species inhabiting my bioregion—but that will be a joyful day indeed. For several decades the entire U.S. population of this subspecies of the gray wolf has been confined within fences. Captive breeding has been successful enough to risk moving a small group of animals back into the wild.

In my other bioregion, the Hudson River Estuary, a much-loved endangered species is being wooed back to health in this most urban of urban landscapes. By the mid 1970s, the peregrine falcon population in the forty-eight contiguous states had been reduced to fewer than forty nesting pairs—none east of the Mississippi River. The pesticide DDT, concentrated upward through the food chain, was responsible for the loss. DDT was banned domestically; later, more than a thousand captive-bred peregrines were released. In New York City this species has made an astounding comeback. The city boasts a dozen nesting pairs on its bridges and tall buildings—which, to a peregrine, resemble cliffs that traditionally served as nest sites. The city is flush with food: pigeons and starlings galore.

Some bioregions are home to more than endangered species; they contain great groups of endangered taxa. Here in the various bioregions of New Mexico, one out of seven fish species that were alive when Coronado arrived are now extinct or extirpated (that is, extinct in this state). A third of those that remain are threatened or endangered. Meanwhile, a whole class of organisms is imperiled throughout the world; amphibians are in steep decline almost everywhere, the cause or causes still mysterious. Whatever one's bioregion, therefore, amphibians merit special attention.

When we think of biodiversity, we tend to image hoards of beetles in Amazonian rainforests. But the entire world looks to the United States for protective management of fully a third of the global inventory of mussel species. Mussels are clamlike bivalves that live mostly in freshwater. But unlike clams, mussels spend part of their unshelled youth attached to the gills or bodies of fish. Two-thirds of the nearly three hundred mussel species native to U.S. waters are at risk of extinction. More than half of these are found in the rivers of Alabama.[7] Bioregionalists living within the watersheds of the Mobile, Tennessee, and Apalachicola rivers therefore bear an inordinate responsibility to Phylum Mollusca and thus to the biodiversity needs of the whole planet.

Another way to approach the biodiversity crisis from a bioregional perspective is to draw attention to "hotspots" of endangered endemic species. In the United States the Hawaiian Islands, Florida, southern Appalachia, and southern California are widely regarded as hotspots for the sheer numbers of species that reside in these places and nowhere else. Less spectacularly, but more thoroughly, one might begin to protect nature's legacy by directing efforts toward the most imperiled ecosystems. Some ecosystems that have been reduced to remnants are homes to many species likewise at risk, but some ecosystems are inherently species-poor and have thus been overlooked. In 1995 Reed Noss and colleagues developed a list of endangered ecosystems—species-poor as well as species-rich—for the forty-eight states.[8] Perusing that list, I learned that I have been dabbling in restoration of a few acres of one such endangered ecosystem, which Noss calls Southwestern Riparian Forests. I have been planting willow cuttings and cottonwoods along the banks and floodplain of a small segment of the Gila River.

I shared the news of my new-found responsibility with my brother in Michigan, and told him that he, too, is responsible for not one but two types of endangered ecosystems. One is a grove of about a dozen trees that has entranced us since we were children. This fragment of Old-Growth White Pine Forest escaped logging in the 1880s, presumably because the trees were too young, and they survived the terrible fires that followed in the footsteps of the loggers (and that turned the topsoils into ash, and thence windblown dust) only because these pines grow on an island of dry ground encircled by a broad moat of swamp and marsh. The other endangered ecosystem begins at one end of the pine grove. It constitutes thirty acres of Tamarack Swamp, whose modest trees rise out of a thick and luminously green carpet of sphagnum moss. By assuming responsibility for something of immense value in our home bioregions, my brother and I have gained a steady sense of purpose that transcends the ups and downs of our domestic lives and careers. Restoring a riparian forest, protecting a grove of trees from chainsaw or a swamp from draining: These callings can be immensely satisfying.

Keystones, Aliens, and Ghosts

The most obvious, and certainly among the most important, inhabitants of a bioregion are the dominant species. For example, two species of cottonwood tree and several species of willow shrub are dominant plants in healthy riparian forest of the Upper Gila Watershed. They provide the vegetative scaffolding on which many species of insects, birds, and other animals depend. Their roots bind soils against the erosional forces that come with the floods. Tamarack and sphagnum are dominant species in a tamarack swamp, redwoods in a redwood forest, sawgrass in the Everglades, kelp in a coastal kelp "forest."

Also of great ecological importance are so-called keystone species. A keystone species is one whose impact on its community or ecosystem is not only large but disproportionately large relative to its abundance or biomass.[9] Like the keystone of an arch, removal of a keystone species jeopardizes the integrity of the whole. With the loss of a keystone species, a cascade of adjustments must take place.

The riparian forest that I tend is home to, and tremendously shaped by, a particular mammal that serves as keystone. For several years this stretch of forest has been intensively managed by a family of beavers, *Castor canadensis*. Beavers are famous as America's foremost ecosystem engineers. Because they eat the living cambium of young trunks and branches of cottonwood and willow, beavers are voracious loggers. In this particular stretch of river, for example, they tend to wipe out the entire streamside forest in four or five years of occupancy. They then move on to greener pastures, returning a dozen years later to harvest the saplings that have since resprouted from the old cottonwood and willow roots. A ten-year-old cottonwood tree may thus be growing from a rootstock many decades old.

Twenty years ago a monstrous flood of a size estimated to occur only once every five hundred years whisked away most of the cottonwood forest (including rootstock) along the upper Gila River. The erosive currents also dropped the river bottom, and hence the water table, making it difficult for water-dependent trees to reestablish on the old terraces that remained. Lush cottonwood forests are therefore few and far between in this stretch of river. For this reason, I have made myself into a keystone organism, working in counterpoint to the beavers. I have ringed a scattering of saplings with four-foot-high chickenwire to give these trees a chance to become sprawling giants. On sandy terraces too dry for seedlings to take root, I have planted young trees and faithfully watered them through the first dry season. Into the raw banks I have plunged willow cuttings, which can sprout roots from woody stems.

Beavers are called ecosystem engineers not only because they keep streamside forests in early stages of succession but also because they build dams that raise water tables, take the force out of floods,

KEYSTONE SPECIES. A representative of one keystone species presenting the accomplishments of another on the upper Gila River.

and reroute river channels. Two winters ago our family of beavers built a showpiece of a dam out of wood and stone and mud that spanned the entire river. Several other beaver families upstream did the same. That winter was unusually dry and there was no spring flood, so all the dams held into the autumn. But eventually a roaring flood took them all out, and none of the beaver families has yet attempted to rebuild their waterworks. The beavers of the upper Gila River are thus fine loggers but frustrated engineers. Half of their keystone function is on hold. But that's not the way it used to be. Old-timers in the area talk of the days when the river was impassable on horseback because of all the beaver dams and ponds and thickets of willow. But the 500-year flood created a problem because no dam can survive even a normal-size annual flood unless a lot of dams and thickets upstream are also in place. Yet the still-meager supply of cottonwoods and willows forestalls expansion of the remnant population of dam builders. It may therefore take a long time for this ecosystem to return to its former state.

What are other keystone species? The first identified keystone was a starfish of the Pacific Northwest whose presence ensured a wide variety of intertidal life clinging to rocky shorelines. When the starfish was artificially removed in order to test how the ecosystem would respond, the mussel population exploded, transforming the once-rich community into a near monoculture of bivalves. The starfish is thus a keystone by virtue of its selective predation of mussels.[10] For coastal kelp forests the sea otter is a keystone because it preys heavily on sea urchins, which, in turn, graze on kelp. In the absence of otters, sea urchin populations swell, and so kelp forests diminish or even disappear.

Fruit bats may be keystones of tropical forests on islands of the South Pacific, because these nectar-feeding mammals are crucial pollinators of many species of flowering trees. Hornbills are keystones of Indonesian forests because these very large birds ingest fruits whole, dispersing over vast distances big seeds through regurgitation and small seeds through droppings. Woodpeckers may be keystones in forests throughout much of the world, as they excavate holes in tree trunks that other birds and small mammals then adopt as nest sites. Prairie dogs and badgers are keystones for excavating holes and building mounds in prairies. Alligators are keystones in Florida's "river of grass"—the Everglades—as they excavate troughs during the dry season to track the declining water table, thus ensuring a water supply for many other species as well. Elephants serve similar functions as excavators during African droughts.

Not all keystones are large animals. Pathogens may be tremendously important in holding populations in check. Before meeting their match in modern medicine, the microbes responsible for various kinds of plague and dysentery ensured that human populations swelling with the onset of agriculture would periodically be pruned. Humans, of course, are now a keystone species for the impoverished ecosystems that do very well by our presence: notably, those in which rats and cockroaches, pigeons and starlings, and various weedy plants dominate. These and other species commensal with humans resemble us in one very important feature: They are virtually ubiquitous. But they weren't so before we provided the transportation to new lands and hacked away at native inhabitants. Humans, rats, and weedy plants are alien invaders of most bioregions in which they are now found. They and we are all exotic species, running wild in bioregions that have no experience in holding us in check.

To know one's bioregion is to discern alien from native. Aliens to North America include ailanthus, Japanese honeysuckle, dandelions, crabgrasses, timothy, salt-spray rose, ryegrass, plantains, Queen Anne's lace, white clover, red clover, and various species of eucalyptus. We sent over to Europe goldenrod; to Japan and China went box elder; Australia is now plagued with our prickly pear cactus.[11] The Africanized honeybee, much in the news today for its recent crossing into the United States from Mexico and for its reputation of fiercely defending the hive, is supplanting not a native bee but an earlier import: the European honeybee.

Zebra mussels and sea lamprey in the Great Lakes, the chestnut blight fungus in the Appalachians: these count among the unintended introductions. But some now-troublesome exotics were brought here and loosed deliberately—and with the best intentions. The water hyacinth, which clogs Florida's lakes and canals, was imported from Central America as an ornamental. Salt cedar (tamarisk), which replaces willow along many rivers of the western states, was brought here from Eurasia for erosion control. Similarly, and just fifty years ago, the U.S.

Soil Conservation Service actually paid farmers to plant kudzu vines to prevent soil loss.

Bullfrogs native to the southeastern states have been introduced into many rivers of the American Southwest and are now ousting the native leopard frogs and stressing populations of small garter snakes. In the eight years I have called the Upper Gila home, I have never seen the native (and vanishing) Chiricahua leopard frog. But there are bullfrogs in abundance. I now count among my friends the very people who, anticipating frog legs dinners, introduced the bullfrog here three decades ago. Nothing in our education prepares us to have the slightest unease about *adding* something to an ecosystem. No alarm bell rings in our heads when we think about planting an alien species in our yard, adding frogs or fish to a favorite pond. Good people do this all the time—people who would never dream of doing any harm to the countryside they love.

But, then, what is native? And are all aliens exclusively bad? Filter-feeding zebra mussels that evolved in Eurasia have taken a toll on native species in the Great Lakes, but they have so thoroughly cleaned up Lake Erie that aquatic eel grass—widely considered a sign of ecosystem health—once again thrives near shore. San Francisco Bay probably holds the world record for invasions by exotic species. Every dozen weeks or so, a new species of aquatic animal, alga, or microbe takes up residence in the bay, having hitched a ride in the ballast water of ships arriving from all over the world. The problem is that the invaders tend to be tough and far from endangered in their own homelands, yet they may edge out native species in their colonial outposts. San Francisco Bay may experience a net increase in biodiversity, but the worldwide inventory of life forms suffers a loss.

It makes sense to do battle with some invaders. Salt cedar is still very rare along this stretch of the Gila; even as a lone individual, I could (assuming the Forest Service would so permit) probably eliminate it in a few weeks of effort. But the bullfrogs are here to stay. The great blue herons and the bigger snakes that prey on bullfrogs have accepted them as members of this bioregion. In a generation or so, we humans will too. The whole point of bioregionalism, after all, is that one can choose a homeland, put down roots, attempt to live in tune with the landscape, and thereby become native. Half my genes have been in North America for less than a century, the other half for two hundred years. Even a full-blooded Native American has been native for only perhaps twelve thousand years. Grizzly bears walked across the Bering Land Bridge from Asia not much earlier. Some three million years ago, North and South America merged, and porcupines waddled northward across the Isthmus of Panama. Horses and camels got their start in North America, but now they are gone. Would loosing Eurasian strains upon the landscape be a restorative action? These are difficult questions. But to even pose them, we depend on paleontologists and paleoecologists to tell us ghost stories.

One of the best tellers of ghost stories is paleoecologist Paul Martin. "In the shadows along the trail," he writes, "I keep an eye out for ghosts, the beasts of the Ice Age. What is the purpose of the thorns on the mesquites in my backyard in Tucson? Why do they and honey locusts have sugary pods so attractive to livestock? Whose foot is devil's claw intended to intercept? Such musings add magic to a walk and may help to liberate us from tunnel vision, the hubris of the present, the misleading notion that nature is self-evident."[12] Such musings can also be deeply troubling. They suggest that even our most pristine, wildest areas are biologically impoverished.

For those of us who have learned to read the signature of ghosts in coevolved plants that have outlived the giant herbivores, the thorns of mesquite in the western states and of hawthorn in the east whisper of a time not long ago when broad-mouthed mastodons and giant ground sloths roamed North America. To know my bioregion is to be aware of these and other ghosts of the Pleistocene. I decided to call Paul Martin to learn exactly who was here in the Upper Gila Watershed twelve thousand years ago when the first peoples spread throughout the continent. Martin confirmed that mammoths and perhaps a few mastodons would have traversed the very mountains and canyons that I now delight in exploring. Although formally designated as wilderness, these landscapes now nourish nothing bigger than reintroduced elk and domestic livestock.

"Would those mammoths have been sculpting the oaks and juniper trees?" I ask Martin. "Would they have pruned the tall vegetation, as elephants now do in Africa?"

"More likely they would have hung out around the floodplain, eating cottonwoods. We do know that mammoths ventured up to timberline. The carcass of one individual has been found in the Wasatch Mountains of Utah at elevations that would have been nearly alpine twelve thousand years ago. That particular animal may have died of starvation, as its preserved gut contained mostly mud with some fir needles. But mammoth dung deposits in the Bechan Cave of Utah mainly contain coarse graminoids, grasses, and some sedges. Laboratory analyses indicate mammoths also ate salt bush and even some blue spruce."

"So when I'm out hiking around here," I surmise, "I can imagine these mountains and canyons bearing mammoths and mastodons. Who was preying on them?"

"Certainly you would have had dire wolves there," he answers. "We don't have any evidence in Arizona of either of the two species of sabertooth cat, but *Smilodon* shows up in the fossil record of New Mexico."

"Sabertooths are so bulky around the shoulders," I comment. "I bet they would have taken prey by surprise attack in a forested or rocky landscape, as we have here, rather than trying to run down something out in the open."

"Yes. Out on the plains there would have been the American cheetah and the American lion—the lion almost exactly the same as its African counterpart."

Martin's mention of cheetah triggers a memory of a recent article I had clipped. Just a few months earlier, *The New York Times* ran a prominent story about the fleetness of the American pronghorn.[13] The question posed was why the pronghorn is able to hit sixty miles per hour when no predator can come close to that speed. The answer is that the pronghorn is running from a ghost predator. It was adapted to outrun the American cheetah, and it hasn't yet shed that adaptation.

Finally, I ask Martin about my favorite extinct organism.

"There were four genera of ground sloths in North America, and each had a single species." He continues, "The one I'm most familiar with, from dung and remains in caves of the Grand Canyon, is likely the same sloth that would have been in your area: the Shasta ground sloth. It's the smallest of the four—not much bigger than a black bear. The biggest one of all lived in South America, Central America, and the southernmost United States from eastern Texas to Florida.

"The Smithsonian displays an awesome skeleton of that giant," I enthuse.

"Yes," Martin adds, "it's over fifteen feet tall, sitting up. The museum in La Plata, Argentina, has fabulous reconstructions of that and other sloths. I remember walking around in a trance looking at them."

"There's something about giant ground sloths that I find so attractive," I say, "so mysterious. More than any other extinct animal, ground sloths are on my mind when I am out on a hike. I guess I

SKELETON OF A GIANT GROUND SLOTH in the Smithsonian Natural History Museum, Washington, D.C. (Photo by Chip Clark, no. 76-5050, 17A; reprinted with permission of the Smithsonian Institution)

mourn for them. They were such strange creatures. How did they get around with having to walk on the sides of their feet?"

"They had to be slow. It's just impossible anatomically to get any speed out of them. With that turned-over, rotated-over foot, plus their great bulk, they have no appearance of swiftness."

"So how did they escape predators?" I ask.

"My imagination is all I have to go on, but I think they could have thwarted an attack by any of the large predators. If a sabertooth cat leaped on a ground sloth from behind, the sloth could just reach around with those big, long arms—with claws three times longer than that of an African lion—and simply rake the attacker in two."

"Yes, and as an edentate [and thus a relative of armadillos], their skin was very tough," I add.

Martin confirms, "In the mylodontids [one family of ground sloths] the skin was fortified by pebbles, or ossicles, of bone."

I shift the topic to eating habits, recounting how, on the long drive to Albuquerque, I conjure up an image of ground sloths browsing in the distance on a vast plain that is a virtual monoculture of creosote bush. No big animal in North America today will touch the stuff. Without the sloths, there is a gaping hole in the ecosystem, I surmise.

Martin corrects my story. "At one time I thought ground sloths were eating creosote bush because we found pollen from that plant in their dung, but now I'm not so sure. Most of the sloth dung is globe mallow and mormon tea."

"So nobody was eating creosote bush?"

"Camels," he offers. "Camels might have been eating creosote bush. Before the Civil War, Lieutenant Beal caravanned camels across New Mexico and on to California, as part of a military experiment. He reported that the camels ate creosote bush. I fed some once to camels in a zoo, and they seemed happy to munch on it."

Several more extinct animals enter our conversation: brush oxen in Utah, a small mountain goat in the Grand Canyon, and an enormous armored glyptodont—an edentate that was almost as big as a Volkswagen Beetle—that spread out of South America as far north as Sonora, Mexico. "You can identify glyptodont genera," Martin advises, "by noting whether they have a club or a spiked mace at the end of the tail." Many more kinds of magnificent mammals that were in North America when the first humans arrived are mentioned in Martin's books and articles.[14] These include elephant-like gomphotheres, giant peccaries, several bison species, a four-horned antelope, a giant armadillo, a bear bigger than a grizzly, and a beaver as big as a black bear that probably did not fell trees and whose remains are found in the Midwest.

Paul Martin is principal architect of the Overkill Hypothesis, which accounts for these recent debilitating extinctions. Put forth in the mid 1960s, the Overkill Hypothesis is now so widely entertained that it earned a place on an interpretive sign for the Pleistocene mammal exhibit of the American Museum of Natural History in New York City. Paleo Indians are depicted on the sign, along with this provocative question, "Were they responsible for the extinction of the ground sloths?" As mentioned in the previous chapter, I prefer to substitute *we* for *they*. Ethnicity is irrelevant here. Humans of every stripe brought about the same sorts of destruction in all lands in which we suddenly arrived as spear-toting invaders. This means on all continents except our land of origin: Africa. Geronimo's ancestors may have wiped out the megafauna of North America; mine surely slaughtered the European contingent.

I read Martin a passage from one of his works: "To behold the Grand Canyon without thoughts of its ancient condors, sloths, and goats is to be half blind."[15] This was written four years before the California condor, an endangered species of carrion feeder with a wingspan greater than that of any other bird in North America, was reintroduced into one of its Pleistocene haunts: the Grand Canyon. The reintroduction took place just three months before our conversation. "Do you think they have a chance to survive there?" I ask.

"That's a really good question," Martin responds. "I don't know."

The reason we both worry about the fate of the condor in the wild is that a "trophic cascade" rather than overhunting probably extirpated this magnificent vulture—along with other big carrion-feeding birds, including a stork. Like the carnivores who depended on the large herbivores, the carrion feeders would have starved in their absence. Condors held on in coastal California, where beached whales and other marine mammals offered a reliable food supply.[16] But Paul and I are both thrilled with the prospect that restoration of this remnant of the Pleistocene is at least being attempted.

"If we can just grab roadkill deer and elk and throw the carcasses out there for the condors . . ." I begin.

"One of my stranger thoughts," he interjects, "is that all of Ed Abbey's friends dedicated to retaining wild animals, and who don't believe in cemeteries and burials, should consider consecrating their carcasses at death to the condors."

"That's not such a strange thought," I contend. "I've dreamed about doing that for our local vultures. Summer mornings more than twenty turkey vultures roost in nearby cottonwoods, then fly up to a couple of dead pines at the top of the driveway, where they spread their wings to catch the heat of the sun as it tops over the canyon. An hour or two later, I get to watch them catch the thermals rising from the east-facing cliff. And so, the idea that upon death part of me could be up there soaring is something that crosses my mind quite often. I find it comforting, actually."

We both laugh, and he continues, "The condor and wolf reintroductions are breathtaking, Connie, but we have a long way to go. If we truly want to reconstruct the Wild West, we've got to venture beyond the living fauna of North America. We must borrow from other continents. At least we should consider it. Yes, wolves are fine and condors are fine, and the other living large animals like elk and bison, mountain goat and mountain sheep, are all great. We need to extend their restricted ranges. But even with all those mammals, we are still working with only a third of what was here before the losses began. It's amazing how many proxy species are reasonably close to the vanished North American megafauna, and some are in serious need of help on other continents. The African lion, for example, is an exceedingly close relative of the lion that appears here in the fossil record of the Pleistocene."

"We'd have to provide some large herbivores for the lions to prey on," I venture, rather skeptical of his bold idea.

"True, it would take a Turner-type wallet to have a go at this level of reconstruction. We'd need to go in for bison in a big way, but it would take more. We'd need proxies for mammoths, and we'd need to bring back camels and horses. Those are the big four."

I can easily guess the identity of the mammoth proxy, but I wonder, "Who could possibly fill in for the giant ground sloths?"

"The ground sloth is the toughest of all," he responds, "because that order [Edentata] really took a hit as far as large animals are concerned. We can't go very far in reintroducing from South America the biggest living edentate, the giant anteater. Ecologically, though, I think a rhinoceros might be a fair equivalent. The fossil record does show them in North America in the mid Tertiary. Rhinos were here at one time, so it wouldn't be totally absurd paleontologically to think about bringing them back."

The conversation drifts to other topics. When it ends, I am left to ponder Paul Martin's audacious proposal. Lions and cheetahs and horses and camels and elephants—*maybe*, I conclude. But nothing short of a ground sloth will do for ground sloth. Rather than settling for rhino proxies, I prefer to hold out for Jurassic Park–style biotechnology to resurrect my favorite ghost. Head spinning, I take another look at Martin's 1992 essay. It closes with this moving statement: "This, then, is our birthright, a continent whose wilderness once echoed to the thunder of many mighty beasts, a fauna that eclipsed all that remains, including the wild animals of Yellowstone and Denali. The mesquites and the honey locusts remember how it was. The rest of us are learning."

POLARITIES

The science of ecology provides a great deal of understanding about ecosystems, which can enhance our awe and reverence for the natural world wherever we may live. By way of science, for example, we learn of the irreplaceable services performed by healthy ecosystems: purification of air and water, flood control, retention of topsoils, nutrient cycling, even the recycling of human-generated wastes—provided the volume doesn't overwhelm the system's ability to render the wastes harmless. The gifts nature affords so freely can, however, be costly to restore. It will cost the federal and state governments billions of dollars to restore the Florida Everglades to some vestige of its former glory. Canals must be undone, curves returned to channeled waterways, lands and waters reclaimed from agriculture. Only then will the pressure of seaward-seeping surface and ground waters be sufficient to push back landward incursions of saltwater. Only then will the sawgrass expanses woo back alligator, woodibis, and turtles in abundance.

There is a tension in the bioregional movement between the impetus to preserve and recover the near-pristine and the impetus to restore the ravaged. Surely we need to do both, but difficult choices about priorities must sometimes be made. For example, do we channel limited public funds to preserve endangered species or to clean up toxic dumps that primarily endanger human health? In the case of the Florida Everglades, a restored ecosystem would serve both human and nonhuman needs, because the landward creep of saltwater threatens the supply of freshwater on which cities in south Florida depend. Sometimes a decision to preserve the near-pristine raises difficult questions about the appropriate degree of intervention. For example, the decision to save the endangered California condor by capturing the remaining wild individuals and breeding them in captivity was controversial. Captive-bred condors released back into the wild still carry a reminder of their captivity. They are equipped with radio transmitters to allow tracking their whereabouts. As for landscapes, even the most untouched may call for a degree of careful intrusion because they may, in fact, have been disturbed by us at some time in the past.

Recall the idea in the previous chapter that islands are everywhere. Landscapes are fragmented; nature reserves are surrounded and isolated by civilization. Yet no place is truly an island, and that means that no place is truly pristine. Human refugees displaced by war or famine, along with impoverished peoples everywhere, slip into nature reserves and extract whatever they desperately need. Even in wealthy countries, nature reserves suffer if they are downwind or downstream of urban centers or industrial agriculture. One community's effluent is an endangered species' drinking water. Acid rain knows no national borders, and alien species do not halt at posted wilderness boundaries.

The reverse is true, as well. Migratory species spend part of the year beyond the bounds and protection of a single bioregion. Songbirds that sing and breed in the temperate zones and higher latitudes must head toward warmer climes when frosts put an end to insects. America's much-loved warblers of the eastern forests are vulnerable to habitat loss and other hazards throughout Latin America. For example, the American redstart winters in the Caribbean, the Tennessee warbler in Mexico, the bay-breasted warbler in the northern Andes, the blackpoll warbler in the Amazon. From the perspective of bird lovers in the tropics who regard these same species as their own, the worrying begins when the birds fly north to breed in a highly fragmented, industrial landscape.

Islands are everywhere, yet no place is an island. Our feathered and foliaged kin suffer the paradox.

Another polarity concerns the degree of management. To intervene or not to intervene: That is the question.[17] Here in the Gila Wilderness, as in forested wilderness areas throughout the dry western states, protection of natural processes is complicated by the fact that active management in the past—notably, fire suppression—has created a far from natural situation. The forests are bloated with down timber, resting on a thick carpet of organic duff. They are dangerously combustible. Unnaturally intense catastrophic wildfires would ensue without continuing human intervention. The strategy now is to initiate controlled burns in patches of forest outside designated wilderness, and to do so during the late fall and winter when cool temperatures and moist conditions reduce the risks of escalation. In the wilderness, however, the strategy is simply to allow some fires to burn—whenever and wherever nature chooses to kindle them. Unfortunately, fires occur naturally only during the summer season of intense lightning storms, which begins at the hottest and driest time of year.

Similarly, how much and what kind of intervention should be undertaken on behalf of endangered species? As noted in the previous chapter, the Gila trout is endangered in stretches of the Upper Gila Watershed that are managed as designated wilderness. Yet a hands-off policy would be their doom because exotic trout species now swim the same streams. Gila trout can survive the competition and the temptation to interbreed with strangers only in isolated tributaries in which a waterfall blocks the upstream movement of alien fish. One such tributary was fortified with a small concrete dam two decades ago—which means a dam was built in the wilderness. Similarly, to reduce the impacts of cattle grazing under private permit, the wilderness has long been parceled into pastures by barbed wire fence, and recently the Forest Service has entertained the idea of building more rainwater catchments in the dry uplands in order to lure cattle away from the thin and fragile threads of river vegetation. Overall, how does one manage a so-called wilderness given these intrusions? How should public land managers balance competing demands? A scientific understanding of the way the ecosystem works is essential for making such judgments, but far more than science is involved, too.

By way of science we learn of another polarity: stability and change. Both are integral and natural players in all bioregions. There is a balance of nature, but there are also natural imbalances that sweep through ecological systems from time to time. Until a decade or so ago, the balance-of-nature worldview prevailed among ecologists—as it still does in popular understanding. Now the imbalances of nature have captured the scientific spotlight. The "shifting mosaic," or "flux of nature," paradigm is on the ascent.[18] Ecologist Daniel Botkin contends, "Nature undisturbed is not constant in form, structure, or proportion, but changes at every scale of time and space. The old idea of a static landscape, like a single musical chord sounded forever, must be abandoned, for such a landscape never existed except in our imagination."[19]

Accordingly, it is no longer acceptable to attempt to hold a particular patch of ground in what George Wuerthner calls "an ecological time warp."[20] Provided an area is vast enough, there will always be climax vegetational communities within it—but exactly where those communities are to be found will shift from century to century. For example, a shifting mosaic of full-canopied cottonwood forest, early successional vegetation, and everything in between will dance up and down the upper Gila River through time, owing to the vagaries of floods, downed giants that redirect water flow, and the comings and goings of beavers. Ecology thus teaches us to see beauty both in the building up and in the washing away of river deposits, both in the growth of a forest and in its demise by floods and fires and incisors.

Stability and change: Both processes are natural, both are necessary. Scientists trained in wildlife management once assumed that a steady mix of predators (or hunters) and prey was an ideal shared by conservationists and nature alike. But there is now suspicion that cyclical crashes may be crucial to long-term ecosystem health. Artificially imposed stability may be detrimental over the long term. Only when herbivores are severely depressed might seedlings preferred by browsers reach the sapling stage. But, again, how do we know we have hit upon the truth even now? Most frightening is our new awareness of *nonlinearities* in dynamic systems, such as ecosystems. This means that although it may take only one more unit of stress or abuse to coax a system into a wholly different ecological regime, it might require ten or even a hundred units of restorative care and a great deal of time to nurture the system back to a traditional or preferred state. Notably, when all the topsoil is flushed downstream or blown into the heavens, it could take centuries, even millennia, for the land to heal.

Hands-off wilderness management allows us the luxury of being wrong. If we can resist the hubris of thinking we know what is natural, or know what is best, then benign neglect may be the ideal form of management for nature reserves that are big enough and pristine enough to handle their own affairs. The directive for wilderness "managers" is thus to err on the side of underinterference. What will be, will be. After all, nowhere besides wilderness can one honor the mysteries on which all biodiversity—past, present, and future—depends. These are the processes of evolution, about which we know a great deal in the abstract, yet little in practice.

As Dave Foreman puts it, "Wilderness is the arena of evolution. Wilderness is the real world, the flow of life, the repository of three and a half billion years of shared travel."[21] Foreman was one of the founders of the Earth First! movement and is coinstigator of the visionary Wildlands Project. The Wildlands Project, with its century-plus time horizon, seeks not just to protect the sparse fragments of wild lands that remain in North America but also to actively recover tampered-with lands and nudge these toward a return to self-maintaining natural processes.[22] It is a bold and—depending on your viewpoint—inspiring, crazy, or sick vision of the future.

"Wilderness areas, more than anything else, challenge us to be better people," writes Foreman. "Wilderness areas are the quiet acknowledgment that we are not gods. Wilderness asks: Can we show self-restraint to leave some places alone? Can we consciously choose to share the land with those species who do not tolerate us well? Can we develop the generosity of spirit, the greatness of heart, not to be everywhere?"[23]

Can we indeed develop the generosity of spirit not to be everywhere? One way to proceed is by way of science—to know and celebrate the facts of life in a green context, but to go deeper, too. We each need to feel the pull of the pageant of life and our own enormous responsibility for the continuance of the Cenozoic phase of the evolutionary epic in as much of its grandeur as yet remains. We need to cultivate capacities to see cherished kin in every nook and cranny of the land and sea, whether sliming slowly across an intertidal rock or clinging to a mountain cliff. We can begin to do these things by setting root, with loving devotion, to a given, chosen, or simply stumbled-upon home.

A Conversation with Hands-Off Stewards

It is easy to love the Gila. The heroes are those who can love the Bronx. That's why I sought out bioregionalists in the Hudson River Estuary, rather than the Upper Gila Watershed, to engage in conversation. Paul and Julie Mankiewicz are both biologists. Each grew up within about thirty miles of where they now live, and they are among the prime movers for the ecological restoration and bioremediation of abused wetlands in and around New York City. If you are a bioregionalist in New York City, you are by definition a restorationist.

I begin by telling Paul and Julie about an essay by the writer and naturalist Terry Tempest Williams.[24] On a trip to New York City, this native of Utah and celebrant of Utah's wild areas visited Pelham Bay Park, near where the Mankiewiczes live, but Williams did not find the park at all to her liking. She did see herons, but the cattails were tattered, the backdrop concrete, the water tainted with oil slicks and the smell of sewage.

"To top it off," I tell them, "she sees a dead dog washed up on the beach."

"She sees a dead dog; I see a copepod environment!" Paul exclaims. "Oh, those copepods are going to love it, with all those amino acids coming off."

Paul has a way of being funny but dead serious at the same time. His ability to see not the carcass but the small arthropods that find it a palace is exactly the sort of utilitarian aesthetic that restorationists must cultivate in order to see glimmers of hope in a ravaged ecosystem.

"Did she mention what kind of dog?" he asks.

"A Labrador," I say.

"That's especially good. They last a while. Not one of those shot-in-the-dark Pekingese that last maybe three, five days."

When the laughter subsides, Julie enters the conversation. "I'm actually stunned by her reaction," she says. "Pelham Bay is astonishingly beautiful."

"So you don't feel deprived living in the Bronx instead of, say, Utah or the Sierras?" I ask.

"It's true there aren't a lot of charismatic creatures here," Julie says, "but there are coyote in the park, and there certainly are foxes. It's one of the major flyways for hawks."

"—and monarchs," I add.

"I remember a trip we took to the Olympic Peninsula," Julie continues. "Yes, the Hoh River valley is gorgeous, but you take two steps outside the designated park and it's all clearcut. Any wilderness is an island. Yosemite is a pretty big island, but it's an island. What is pristine wilderness? There's nowhere."

I join in. "Because New York City is so built up already, things can only get better as the community begins to take an interest in bioremediation. Whereas if you're living somewhere that is still very natural, there's always the concern that it will be eaten away—more homes going in, more trees coming out. There's this sense of a dire trajectory downward. Here I never have to worry about bonding with a particular woodland and then watching a bulldozer scrape it away. Central Park is always going to be there—and the parks department is now actively restoring the native vegetation."

"I remember watching a red-tailed hawk bring a pigeon to feed its young around 66th Street near the Arsenal," muses Paul.

"I remember when a pigeon got it right on the steps going down from Cathedral House," Julie says. "It was a peregrine that did it."

"My editor told me he walked into his twentieth-floor office one morning and saw a mangled pigeon head and wings on the window sill," I report, "presumably the work of a peregrine."

"I remember being on the parkway to Kennedy Airport, as we

were going to spend our honeymoon in Sweden at an ecological wastewater treatment conference," Paul says. "Flying by was a peregrine with a starling in its talons."

Paul and Julie Mankiewicz run their own business of bioremediation and ecological education. They also put in a lot of volunteer time on boards and committees dealing with these issues. Both have doctorate degrees, but as Paul explains their choice of work, "You will never do decent theory without applications. What's killed theoretical biology to date and has made ecology an essentially useless field is there are few actual attempts to test theories by restoring and rebuilding damaged landscapes."

"As time goes on," Julie says, "I realize that as much as I love research, I'm also an educator at heart. I think the best way to teach science and to re-engender empathy—you can't teach empathy—is to give people the sense that they can restore something. We need to give people a sense that they can actually make something better through their own efforts. For children, then, learning science becomes something for a purpose—to keep a particular marsh alive or an aquarium of native fish and mussels and crabs that we help install in a school."

She continues, "I remember when we took a troop of Boy Scouts out for some marsh restoration work. This was at an inlet just south of the Pelham Bay landfill. It was a cold April day, and the boys were out there all day planting *Spartina*. The parents really wanted to go home when the planting was done, but we had promised the kids that they could use the microscopes we brought with us to see what was in the mud. And they really wanted to. One later chose an eagle scout project in which he helped us do a census of another planting around the landfill, and now he's in college studying for a degree in marine biology."

"So," I surmise, "kids need to get a sense that they are contributing to something—that they can make a real difference."

"Especially because so much of what they learn in school about the environment makes kids—and everyone—feel rather powerless and hopeless," Julie responds. "We need to help them see that nature is not just vulnerable wilderness that's getting torn up. Nature has its own regenerative capacities, and our own health depends on that. So when you're out there planting marsh grasses, you're not just helping nature but you're also improving the air that we breathe, the waters we swim in."

I met Julie and Paul Mankiewicz soon after I moved to New York City in 1989. At the time of this conversation, I knew a good deal about their restoration work and their teaching experiences but had never delved into matters of spirit. Nevertheless, I suspected they would have some thoughtful answers along such lines, because both were involved in ecological programs at the Cathedral of St. John the Divine. In an alcove of the Cathedral, Paul and Julie installed a huge aquarium of salt marsh creatures native to the Hudson River Estuary and a terrarium of native ferns and mosses. In their home, where this conversation took place, smaller versions of these two installations dominated the cramped, ground-floor living space. Outside, their small yard holds a newly dug pond with two salamanders and hopes for spring peeper frogs. I prompted a turn to matters of spirit by asking each to identify an ecological principle that also had spiritual significance. What principles, in particular, should be taught, communicated, celebrated?

Paul begins. "The dynamic side of ecological systems is key. By this I mean that the slate of organisms that might live at a particular location and the ecological systems themselves are not determinate. We see this in restoration work. A different mycorrhizal relationship, for example, might shift the whole outcome. Over time, some new organism will come along, and by brokering a new relationship with fungi and bacteria, it will change the whole nature of the system. So you can count on the fact that things will change."

"I can see how you would be more comfortable with the principle of dynamic change," I suggest, "than would, say, a conservation biologist trying to secure the survival of an endangered species vulnerable to change. The organisms that concern you are really the gladiators of the ecosystem, like *Spartina* grass, and restoration work, by definition, promotes change in a run-down, damaged system."

Paul agrees. "*Spartina* are hardy plants at the base of the food chain of the North Atlantic. The pipe full of run-off toxins dumping into the inlet that the Boy Scouts worked on was probably not damaging the plants so much as was the physical loss of habitat owing to dredging. I work with the gladiators because if you don't have the base of the food chain, you don't have anything."

"My favorite ecological principle with a spiritual side," says Julie, "is Liebig's Law of the Minimum. Any ecological system has one primary limiting factor: a nutrient, water, something. So, in our restoration work, if we supply that limiting factor, productivity in the system goes up. This means that as long as we are selective in what we do, we really don't need to do very much in order to get great results. I try to abide by this same principle in raising our children. To keep them safe, you only have to watch them at key moments. Children or salt marshes—we just need to look carefully at what they need and supply that. They'll do the rest on their own."

"That's what I call *hands-off stewardship*," says Paul. "I came up with that term while talking with Jeff Golliher, the environmental priest at the Cathedral. It is a way to take the hubris out of the stewardship mandate that Christians see in the Bible."

"Staying on spiritual matters," I say, "what does your understanding of ecology lead you, each of you, to regard as being of ultimate value?"

Paul begins. "The recognition of human beings as ends in themselves is a cultural accomplishment, hard-won and of extreme value. But as yet there's no recognition of the value of the diversity of beings that support each of those individuals and each of those cultures. So, in my view, ultimate value should also be placed in the diversity of beings. You will never be able to maintain an individual with character and capacity without that diversity. The problem with kids' understanding that we talked about earlier is that they have not learned to recognize any set of beings that calls into question their easy assumptions about our own species' uniqueness."

Julie responds, "My immediate reaction is that the purity of air and water is of ultimate value. To me there is something sacred about waterways, rivers, streams, lakes, sounds, oceans. Given that 99 percent of the species that have ever been on this planet are now extinct, I don't think that saving particular species, a given particular diversity, is necessarily always what we need to do. But there is something about the way the watersheds work, the planet works, that needs to be preserved.

"One of the easiest ways to elevate our regard for water" she continues, "is to remember that water is the universal solvent. I actually think about that a lot—trying to figure out what soaps to wash my kids in. Most of the time I just dunk them in water. Water is the universal solvent. So pure water and the natural systems that supply pure water are of the highest value. In my mind, it is an absolute crime that the city of New York would even consider, as it is now, building a massive artificial system to filter the water we direct here from the Catskills, the Delaware, and even the damaged Croton system. To give up what nature has so long given us for free—pure water—and to start doing it ourselves while we let the watersheds deteriorate with human abuse and development—to lose the ecological integrity of the whole region simply because we believe we can replace those ecosystem services—this is the worse sense of hubris. It's a travesty. It's a violation of what is sacred."

"Julie, you use the word *sacred*. How do you feel about the word *religious*?" I ask.

"*Sacred* is a powerful word for me. But *religious* speaks more to a group of people devising rituals. It's more the culture of the sacred."

"So, the sacred is what is intrinsically there," I interpret, "but it is out of the sense of the sacred that one develops religious packaging."

"Which may or may not work for you," she adds. "Christianity has not worked particularly well for me because I think it has not included enough of nature in the Golden Rule. That's the challenge we face at the Cathedral, and that's why we put the exhibit in the Earth Bay. The challenge for all established religions is to take what we've tried to do with the human community and extend that to the rest of the beings on this planet."

"So you get your sense of the sacred directly," I offer, "direct contact and also by way of your scientific understanding—the notion that water is the universal solvent and that vast ecosystems are busy, on their own, purifying water, and will continue to do so in perpetuity if we just let them be. Your sense of the sacred is not mediated by something that might be called religious."

She affirms, and I turn to Paul. "Paul, are you willing to use the word *sacred*?"

He pauses. "Sacred. The real question is, what isn't?" he says.

"But aren't some things more sacred than others?" Julie prods him.

"Hint: H_2O," I whisper.

Paul and Julie Mankiewicz, despite their wariness of religious ritual, do don white gowns once every year and march down the great aisle of the Cathedral of St. John the Divine. The event is the October celebration of St. Francis Day. Francis of Assisi was a thirteenth-century monk who befriended animals and worked for a more nature-friendly interpretation of Christianity. Pet owners who might not step into a church or synagogue of any denomination on other days of the year flock to the Cathedral with their cats, dogs, goldfish, and other animal companions to be part of this bizarre and moving event. Those who wish may have their companion blessed by clergy at the conclusion of the service. The climax, however, is the procession. Up the aisle come a menagerie of creatures—each with a caretaker to keep the animal relaxed and under control. The largest is an elephant. But an eagle, a boa constrictor, a camel, and other beings also become absolutely awesome in context of row upon row of hushed humans in one of the loveliest built spaces on the planet. Paul and Julie are also caretakers of beings in the procession, and they always come at the tail. One will usually carry a plant, though the species changes from year to year. But under no circumstances will the crowd tolerate substitution for what the other one holds. That is always a glass sphere filled with cyanobacteria, also known as blue-green algae.

"When you are carrying the cyano ball in that procession, what do you feel?" I pose this question to Julie, who carries the ball most often.

"I feel a real tension," she responds. "It's a very powerful moment because people are instructed that for their safety and the safety of the large vertebrates, they must be quiet. But it is also something of a comic moment to be walking with the bacteria, because the dean usually talks it up. He talks about the trillions of beings alive in that glass bowl and how they are among the first organisms that lived on this planet, and that they are relatively unchanged from what they were back then. So I hear through the hush kids whispering, 'Here come the algae! Here come the algae!' On the one hand I feel honored, but I wonder if it is an anticlimax for some."

"Are you kidding?" I counter. "I mean, how many people have

THE PROCESSION OF ANIMALS, St. Francis Day 1996, at the Cathedral of St. John the Divine in New York City. (Photo by Martha Cooper, courtesy of the Cathedral of St. John the Divine)

ever seen algae or cyanos in the shape of a sphere. It's perfect. The ultimate symbol of purity, unity. Perfection of color, no texture."

Paul agrees. "Yes. Perfect pigmentation. A perfectly dense blue-green."

"Holiness right there," I conclude, "that more than anything else."

I continue, "So here you two are, participants in what is the most moving church service I have ever come upon. That's a role to die for, in my view. And yet you are not believers in the traditional sense."

"We're Episcopalians because Dean Morton is Episcopalian," Paul explains. (James Parks Morton, often called "the green dean," was, until his retirement in 1997, the head of this very large parish.)

"Does the dean regard you folks as religious?" I ask.

"Actually, I think he regards us as models of religious people," Paul answers.

Julie elaborates, "I think the dean recognizes something of what may be termed a calling in what we do that is very much in line with his aims."

She continues, "When we had our kids christened, Paul wrote a beautiful statement about the meaning of water. We had that read at the christening. The dean had been urging Paul for years to help rewrite parts of the mass. Well, he liked what Paul wrote so well that now it is offered as an alternate liturgy that any parent in this church can choose to have read for their own child's baptism."

"You know," begins Paul, "Christianity supposedly threw out the old law and replaced it with two things: love this, love that. It was supposedly a movement from justice to empathy. But where is the empathy? People don't bother to get to know the organisms in their environment. They don't know the names of trees; they certainly don't know the names of mosses. All insects are bad. People have no empathy for beings beyond themselves.

"Nietzsche was quite right," he concludes. "If you don't have empathy for the beings around you, then certainly God is dead. Since religion is dead, we do biology."

Creating Science-Based Rituals

Here is the text of the christening liturgy written by Paul Mankiewicz and performed by the Very Reverend James Parks Morton in the Cathedral of St. John the Divine, New York City, on St. Francis Day, 1995.

The Baptism of Phoebe Simone Mankiewicz
and Theodore Francis Mankiewicz

Remember that water poured over these two beings will be with them and in them when they too pass away. This water which touches them will circle the globe, giving life to all beings around the Earth. May the circumference of water's reach teach these beings, and all of us, that good and evil, whatever passes through us, reverberates through all creation.

This very water will become, seasons from now, ice in the far north. Most will return to the sea. How much ice, how much sea, depends on us, and how warm we make our earthly home.

Poisons we have added to water inevitably become part of these two beings whom we hope to consecrate with water, as these poisons are part of us. May these and other sins against one another and the very circumstances that support us not blind us, or these children, to the daily gift of water.

As we treat water, we treat our neighbors, our children, and our children's children. Life is born of water, and born again of water in each day's turning of the Earth. May these beings, and those of us who would teach them, meet in water the power of creation, an empathy for all beings, and the spirit which connects each to all, in the name of the Father, of the Son, and of the Holy Ghost.

This celebration of water poured from the heart of a biologist, a very reluctant liturgist. I print it here not so much in the hope that readers might induce their own churches to adopt it (although that would be wonderful). Rather, it is a powerful demonstration of the creativity and capacity for ritualistic modes of writing and speaking that reside in us all. Anyone who takes the effort to learn the ecology of their home bioregion and to cultivate a deep empathy for other beings may be equally inspired to such heights of sacred writing.

In my own church, the Fourth Universalist Society, New York City, I created an opening invocation for Earth Day 1995. It was performed by four women in our congregation, with a few natural props. Here I print two of the four readings. Note that a home-grown liturgy may well include elements of such a local nature that it isn't readily transferable to anywhere else. The science base may be universal but the magic is in the particulars of place.

Earth Day Opening Invocation

In our hearts we call forth gratitude for the air that swirls around and within us, for the ancient microbes that two billion years ago discovered how to crack a molecule of water, feeding on the hydrogen and setting the oxygen free—forever changing the orange-brown sky of primordial Earth into a brilliant blue.

We call forth gratitude for our partners in the cycling of air: the great plant kingdom. Together we make the breathing of the biosphere. With our very next inhalation we will all almost surely receive the gifts of the spring grasses in Central Park; perhaps one of us will receive the gift of a young cedar tree in China that last July drew down the power of the sun, or the gift of a tiny diatom that lived in the Sargasso Sea on the day one of us here was born.

We call forth gratitude for those of our own species who have worked to freshen the air, to free it from the soot of coal burners and apartment-house incinerators. Their good deeds will forever be wafted on the wind. In our hearts, we call forth gratitude.

In our hearts we call forth gratitude for the water that flows around and through us, for the vast oceans that are sink and source of all, for the salty waters in which the animal kingdom got its start 600 million years ago. That same saltiness still courses through our veins. We taste it in our tears.

We call forth gratitude for the freshwaters, too. For the rain and snow that fall upon the Catskill Mountains a hundred miles to the north. Those forested and protected slopes share with us the sweet harvest that spills from our faucets, the envy of urban dwellers from San Diego to Moscow.

We call forth gratitude for the great Hudson River and the sheltered harbors that surround us, a landscape ideal for the waterborne commerce on which we urban dwellers depend. We call forth gratitude for the vast watery realm that bears our bodily wastes, and for those women and men who have dedicated their lives to making our impress on the estuaries and sea less burdensome, and who try so hard to awaken our concern for the myriad beings that live out their lives beyond the shore. The goodness of their deeds will flow with the waters forever. In our hearts, we call forth gratitude.

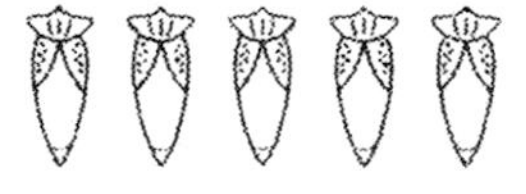

5

Geophysiology and the Revival of Gaia

A meadow in the middle of the sky
A little waterhole in the Vast Space
A dazzling oasis, swirling with blues and whites
A pearl in a sea of mystery

Carl Sagan, Gary Snyder, Diane Ackerman, Edgar Mitchell,[1] and many other scientists and writers (especially astronauts!) have written eloquently on the most powerful image of the twentieth century: the view of Earth from space. The American space program gave us more than an image, however. It prompted a scientific insight that launched a whole new research program. The insight is the Gaia hypothesis. The research program is geophysiology—the physiology of the planet itself, the greatest living system of all.

The image, the insight, and the supporting science have, in combination, revivified a spiritual concept that has long been overlooked by religious lineages that preferentially gaze outward to the heavens or inward to the depths of human consciousness. Something far greater than the human may or may not exist Out There or In Here. But we do know, through science, that praiseworthy greatness can be found All Around. More, we are immersed in the grandeur. Like peas in a soup, we are part of—not apart from—the object of our devotion. We live not *on* planet Earth but *within* Gaia.

> The Gaia hypothesis is for those who like to walk or simply stand and stare, to wonder about the Earth and the life it bears, and to speculate about the consequences of our own presence here. It is an alternative to that pessimistic view which sees nature as a primitive force to be subdued and conquered. It is also an alternative to that equally depressing picture of our planet as a demented spaceship, forever traveling, driverless and purposeless, around an inner circle of the sun.[2]

This and other moving and poetic passages blend with original scientific insights in the book that first set forth the Gaia hypothesis, *Gaia: A New Look at Life on Earth* (1979). Since then, Jim Lovelock—the scientist who originated the idea—has published two more books expressly on Gaia: *The Ages of Gaia* (1988) and *Healing Gaia* (1991). It is the first book, however, that nestles in the hearts of many of the scientifically inclined. We discovered the gaian worldview—the view of Earth as a self-regulating living system—by way of Lovelock's graceful prose and endearing personal narrative.

A serendipitous encounter with that book is what brought me back into science a dozen years after I had taken an undergraduate degree in zoology and moved on to other things. The idea of a living Earth was part of the attraction, but so was the implication that biology is a lot bigger than I had been led to believe in school. Facts I had learned in isolation in chemistry and geology classes now mingled with the facts of life. And I was excited by the prospect that the era of great discovery was far from over. Darwin hadn't done it all. I thus shed the nostalgia picked up from exposure to perhaps too many textbooks and determined to become a science watcher.

One thing led to another. By the spring of 1996, I was attending my third scientific conference on Gaia, this time at Oxford University. Here I had the opportunity to tell the 76-year-old author how much his first book had meant to me.

"That book is very special for me, too," he replied. "I had such fun working on it. I wrote it all in Ireland, which is a very suitable place for such books. I would sit on one of the warm sandstone slabs—part of Hungry Hill—looking out over Bantry Bay into the Atlantic. In some ways the book was a love letter to an imaginary woman. I never dreamt that fourteen years later she would be embodied as Sandy. Had she not read it, we would not have met."

Jim Lovelock met Sandy Orchard at the 1988 Global Forum of Parliamentary and Religious Leaders on Human Survival. She was one of the conference organizers. Having heard a bit about the man and his ideas, Sandy decided to read the book before his presentation. "It had the most profound effect on me. I began reading it during my commutes to London on the train, and I wanted to go up to people and say, 'Look! Have you read this book? It's so important!' I was so taken by it I couldn't wait to meet him." They did meet, and were married soon—and happily ever—after.

Gaia begins with the tale of how this British chemist and inventor came upon the idea. The story opens at the National Aeronautics and Space Administration in California during the early 1960s. This was the time of preparations for the first exploratory mission to Mars, and Jim Lovelock had been brought on as a consultant. Although his formal charge was instrument design, he couldn't help thinking about the broader questions put to the biologists: How can a probe be made to detect life? What sorts of experiments should be done on-site? What kinds of measurements need to be taken?

Lovelock volunteered an answer that was far from welcome, for he concluded that any test would be redundant. There was surely no shortage of pressing scientific reasons to pay Mars a visit, but looking for microbial martians was not one of them. Mars may or may not have been inhabited in the past, but there was surely no life on Mars today. The spectrum of reflected sunlight emitted back through the martian atmosphere and sensed right here on Earth was already evidence of a dead planet. To Lovelock it was all so obvious: All the gases in the martian atmosphere are in chemical equilibrium. There is no life on Mars. Case closed.

To make the connection, to know that an atmosphere lacking reactive gases signifies a dead planet, Lovelock had to go back to the basics. What makes life alive and how does life affect its nonliving surrounds? It is here that he had his first inkling of Gaia—rather, of the idea that would later be named after the earth goddess of the ancient Greeks, at the suggestion of an imaginative neighbor: the novelist William Golding. To begin with, Lovelock understood that life anywhere in the universe almost surely depends on a fluid medium as a source of nutrients and as a sink for wastes. Because there is no ocean on Mars, martian life would utterly depend on the atmosphere for this service. This dependence would work a substantial (and in Lovelock's view, recognizable) shift in the mix of atmospheric gases.

Why? For one thing, living organisms are chemical factories. We all have what is called a metabolism. We eat stuff and we void stuff just to maintain, and we do even more eating and voiding to reproduce. We can't, moreover, all be doing exactly the same thing. If we all ingest molecule A and excrete molecule Z, pretty soon there will be no more A molecules to support anybody, just a useless abundance of Z. Somebody, therefore, has got to find Z palatable, even delicious, and somebody has to excrete substance A. Somebody's waste must be somebody else's food. It thus takes a food web (more a ring than a chain) to make a biosphere, ultimately powered by a star.

Again, because the atmosphere will be used as source and sink, at least a little of everybody's gaseous waste products are going to spend some time in the air before they are captured by the organisms next in line along the web. Some of those waste products are going to be *reducers*—molecules that would love to donate an electron or two—while some will be *oxidizers* that would be happy to accept same. In Earth's atmosphere oxygen is the most prominent of all oxidizers, but there are others too. As for reducers, methane belched by the microbes in the rumens of cows and the muds of rice paddies is ever-present and perhaps even detectable by infrared interferometers vast distances beyond Earth, should anyone out there be curious and capable of looking. When an electron donor and an electron acceptor find one another, a chemical reaction takes place automatically or jolted by a little sunlight. For example, when oxygen and methane meet and shuffle their atoms, their progeny take the form of molecules of water and carbon dioxide. The resulting chemical mix is then at peace; chemical equilibrium prevails.

Until the donors and acceptors all find one another, an atmosphere will remain in a state of unrest. It will contain reactive gases. Because life is ever busy, electron donors and electron acceptors will continuously enter the atmosphere, even as others find partners and consummate their desires. The air will never settle down into chemical equilibrium on a living planet.

This persistent state of chemical unrest is the signature of life. The atmosphere of a living planet would thus, in Lovelock's view, be conspicuously unlike that of a dead planet. A comparison of the atmospheres of Venus, Earth, and Mars (see the accompanying figure) drives this point home. Were life to vanish, moreover, Lovelock and

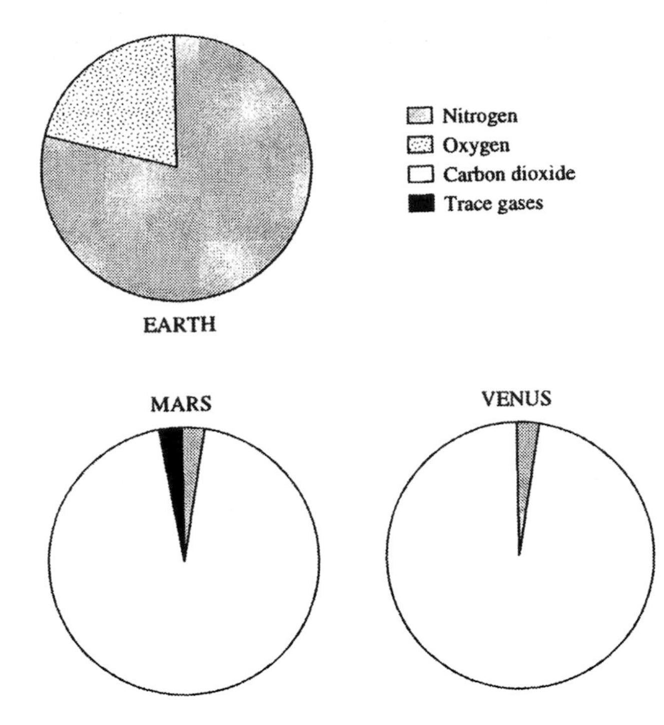

ATMOSPHERES OF EARTH, MARS, AND VENUS. Gases constituting less than 1% of a planetary atmosphere are not depicted.

many other scientists suspect that Earth's atmosphere would eventually come to resemble that of Venus and Mars.

The Gaia insight thus began with a question about Mars that led Jim Lovelock to take a fresh look at our own earthly realm. That fresh look then prompted a bold rethinking of the relationship between life and environment on a global scale. And from this was born perhaps the loveliest of all gaian metaphors. "The atmosphere is not merely a biological product," wrote Lovelock, "but more probably a biological construction: not living, but like a cat's fur, a bird's feathers, or the paper of a wasp's nest, an extension of a living system."[3]

Today, Jim Lovelock's decades-old, heretical idea about how to look for life beyond Earth has been vindicated. In 1996 the idea resurfaced in the guise of a major initiative at NASA. Now that planets in other star systems have been detected (indirectly, however, by the ever-so-slight gravitational wobbles they produce in their stars), we are bursting with curiosity to have a look at them. By building the proper kinds of telescopes and sensors and perhaps even setting them out in space, we might have a chance to do just that. If we can indeed catch an infrared glimpse of an alien planet, the answer to the question "Is there or isn't there?" may be virtually in hand. Again, the spectrum of light reflected from the planet and filtered through its atmosphere serves as the calling card for surface life.

As for Mars, other NASA visits are planned. The exobiology question to be asked this time, however, is one that Lovelock cannot answer by analyzing today's martian atmosphere: Was there *ever* life on Mars? For this, the clues are held captive in the rocks. But the rocks may not all be held captive on Mars.

In the summer of 1996 the science (and even the popular) press was abuzz with fresh speculations about fossil life on Mars. A meteorite picked up in Antarctica in 1984 was the source of the excitement. This meteorite became exceedingly interesting to researchers in 1994, when it was classified as having arrived from Mars, ejected from its home planet when a meteor far bigger struck nearby. Another research team then began to probe the meteorite in detail. This team turned up ancient hydrocarbon and carbonate molecules, a rich store of tiny grains of magnetite indicative of microbial metabolisms on Earth, and tantalizingly ovoid and tubular specks suggestive of earthly bacteria, though the specks were far smaller than any bacteria known on Earth. David McKay and colleagues were careful not to claim proof of martian fossils, but they did argue that the evidence was strongly suggestive of life.[4]

If the meteorite does indeed contain evidence for life, is it a stroke of luck that the small set of known martian meteorites (just twelve) would yield such information? Or did Mars support a teeming biosphere at the time this martian rock is thought to have formed—3.6 billion years ago? Might that teeming biosphere have existed *within* rather than on Mars, as Thomas Gold has proposed?[5] More tenuously, might a martian deep biosphere still exist within porous rock, making use of subsurface water heated to a liquid state by the interior heat of the planet but remaining pretty much isolated from the thin martian atmosphere?

Even just thinking about the possibility that Mars may once have harbored life brings us to another question about the living Earth: Is the persistence of life here for three and a half billion years something that demands explanation? Was it a rough go? And did life itself have anything to do with the awesome record of longevity? Here lies the core of the Gaia hypothesis and the tenet that has generated the greatest dispute—and the most research.

Powers of Persistence

"Life is a planetary-scale phenomenon. There cannot be sparse life on a planet."[6] Jim Lovelock makes this assertion because he thinks the task of planetary management is a big one, requiring a full crew. What, then, are the challenges?

There are broadly two requirements for a habitable Earth. The chemical regime and also the thermal conditions must be kept "tolerable," "conducive," "favorable," or "fit" for life. Although Lovelock now favors these adjectives in explaining the role played by life at the global scale, he initially posited an even more fastidious Gaia, ever busy keeping the planet "optimal" for life. The physiological term *homeostasis* was chosen by Lovelock to depict what constitutes a decently tolerable world. The sum total of life acting together keeps the chemistry and the climate fundamentally the same—homeostatic—from year to year, century to century, and even eon to eon. Life in the aggregate, because it tucks into and manipulates the natural physical processes, ensures such chemostasis and thermostasis by way of *self-regulation,* or active control.

For Lovelock to interest the scientific community in his notion of a self-regulating Earth system, however, he needed to convince his colleagues that chemical and thermal stasis is not otherwise to be expected. Wide and debilitating fluctuations or life-threatening trends, absent a gaian ensemble of life, would have to be seen as the norm. After all, anybody who might suggest that life has a role to play in tweaking Earth's orbit in order to keep this planet precisely circling the sun would be called a crank. Gravity and momentum are fully adequate to the task. In science, Occam's razor reigns. One does not posit a new force or entity when the familiar will suffice. The familiar, in the case of Earth history, would be natural chemical and physical processes, coupled with life's knack for conforming or adapting to environmental changes, by way of the process Darwin proposed more than a century ago: natural selection.

Lovelock's assertion that the atmosphere of a living planet—any living planet—would be wholly unlike the atmosphere of a dead planet was radical only in that it made life far more powerful than was commonly thought. Our own strange atmosphere could be regarded as the natural outcome of a rich veneer of living organisms altering the environment as a by-product of metabolic activity to which the biota then, in turn, must adapt. This is the concept of coevolution. Many scientists have indeed come to regard the huge component of oxygen and the simultaneous presence of reactive gases as evidence of a very active earthly biota. But this evidence for the power of life could not in itself bring skeptics to a fully gaian, regulatory perspective. The postulate of coevolution could be taken as sufficient. Here life is simply regarded as skilled at coping with whatever mess it makes for itself. The theory of coevolution means that there may be no need to posit a self-regulating biosphere that controls atmospheric inputs and outputs in order to keep things favorable for life.

Overall, Lovelock's comparison of the atmosphere of Earth with that of neighboring planets and with a hypothetical lifeless Earth was a compelling argument for the power of life, but not necessarily for the existence of Gaia. This curiosity alone was not enough to catch the attention of colleagues and to entice them to entertain the possibility that a new process might exist, exercised by a hitherto unrecognized entity—something coherent enough to be regarded as a *self* engaged in self-regulation.

Lovelock himself made the leap from a power-of-life perspective to a fully gaian way of thinking when he came upon a problem that astrophysicists had posed: *Why is it that the oceans were not frozen solid at the origin of life?* From the standpoint of astrophysics, thermostasis is not to be expected for a planet orbiting the sun. This is because our star is a member of a class of stars that steadily heat up as they age. The sun was therefore a lot cooler at the origin of life—perhaps 30% less radiant. Given that the average global temperature at the surface of Earth today is about 15°C (60°F), then, all else being equal, the oceans would have been frozen solid three and a half billion years ago. Yet the sedimentary record and the record of life itself indicate otherwise. To solve this "faint young sun paradox," scientists have posited a shift in the chemical makeup of Earth's atmosphere through time. Specifically, Earth's blanket of gases must have trapped a lot more solar heat back then than it does now.[7]

What gas was responsible for the super-greenhouse conditions on the primordial planet? What processes worked its gradual removal, coincident with a brightening sun? Did life have anything to do with that trend? Overall, was the historic shift in atmospheric chemistry that countered the effects of increased solar luminosity just fortuitous? Or is it evidence of a biosphere busily engaged in doing the right thing?

To the question of what the primordial greenhouse gas was, the best answer seems to be the very gas whose greenhouse properties worry us most today: carbon dioxide. At the origin of life, the atmosphere contained perhaps ten thousand times as much carbon dioxide as it holds today.[8] Where did all that carbon dioxide go? It went into the crust, in the solid form of carbonate rock, which includes calcium-rich limestones, $CaCO_3$, and magnesium-rich dolomites, $CaMg(CO_3)_2$. *Why* that drawdown of carbon dioxide took place and *who* or *what* did it is a source of some contention.

Most scientists now agree that a combination of physical and biological processes sucked the carbon dioxide out of the air and sequestered it in rock. Some regard the plodding contributions of physics and chemistry as sufficient, with the biological forcings inventive but superfluous. In this view, the biota simply speeded up what would have happened anyway. There is nothing astonishing about the downward trend in carbon dioxide, they maintain. We are just lucky that geochemistry happens to work in the direction that offsets a brightening sun, rather than doing the opposite.

The underlying geochemistry is a threefold process that begins with the chemical weathering of igneous rocks on the continents, followed by transport of the chemical constituents to the sea, culiminating in their precipitation as carbonates on the continental shelves and ocean floor. Let us begin with the continents. On the early Earth there were no primordial limestones or dolomites. All rock was volcanic. Different volcanic rock types were all variations on a single theme: a lattice of silicate, SiO_4 tetrahedra. Oxygen and silica are, after all, the two most common elements in Earth's crust. The silicate lattice may incorporate iron, magnesium, calcium, potassium, sodium, aluminum, and other positive ions within its crystalline frame. However complicated the chemistries, most silicate rocks do not contain carbon.

Now consider what happens when an atmosphere rich in carbon dioxide pours forth rain onto the continents. Carbon dioxide is highly soluble in water. Soft drinks effervesce upon opening only because the carbon dioxide–infused liquid was packed at a pressure five times greater than the pressure at sea level. Even after the bubbles are gone, the remaining CO_2 (usually in combination with citric acid) keeps the beverage acidic. This is because a molecule of CO_2, when dissolved in water, combines with a molecule of H_2O to form carbonic acid, which then dissociates slightly to form a negatively charged bicarbonate ion $(HCO_3)^-$, leaving a lone, positively charged ion of hydrogen (H^+). Both ions are now free to do mischief, as they search out partners for chemical stability.

This mischief is the breakdown of silicate rocks—without which there would be no soils. The breakdown is a chemical, rather than a physical (erosional), form of weathering. Chemical weathering owing to the acidifying effects of carbon dioxide on rainwater demolishes the connections that bind the positive ions into the silicate lattice. For our purposes, let us follow the fate of just the calcium ions. More and more calcium ions are liberated from the rock and washed down rivers into the sea. This is step two of the solution to the faint young sun paradox. Step three, limestone deposition, is a natural consequence of the fact that water carried to the oceans is continuously recycled to the land by way of evaporation, but calcium is not. It has no gaseous phase at earthly temperatures. Accordingly, as rivers discharge their cargo into the sea, the ions accumulate. Eventually, the concentration of ions becomes so rich that crystals of calcium carbonate (limestones) precipitate naturally, drifting to the bottom. This is what pulls carbon dioxide out of solution (and ultimately out of the atmosphere) and down into Earth's crust.

This threefold process of (1) chemical weathering of silicate rocks, (2) ion transport to the sea, and (3) concentration and precipitation of carbonate is the solution to the faint young sun paradox. Where does life fit into the picture?

David Schwartzman and Tyler Volk contend that life plays a very big role indeed in the sequestering of atmospheric carbon into carbonate rock. Most obviously, life is involved in step three, deposition of limestones. But perhaps surprisingly, life is not *crucially* involved in that step. Shelly limestones do, of course, indicate that life latched onto the calcium and made its own calcium carbonate before the water had a chance to precipitate same; but limestones would have accumulated even in the absence of life. Rather, step one is where life is most important. Life accelerates the geochemical process of chemical weathering, thus rendering it a *bio*geochemical process. In a 1989 publication, "Biotic Enhancement of Weathering and the Habitability of Earth," Schwartzman and Volk estimate that, without life, an abiotic Earth might be as much as 30–45°C (86–113°F) *warmer* than it is today, yielding a global average temperature of 146–173°F.[9]

How did these two scientists reach this shocking conclusion? They calculated the ways in which land life boosts the chemical weathering of silicate rocks. Simple crusts of microbes or lichen that form on rocks produce various acids as metabolites. Rooted plants tend to be more active, producing more acids—some of which may have evolved as adaptations by which plants could access more nutrients held captive in the crystal lattices of silicates. Perhaps most important is that a veneer of life, especially rooted life, allows soils to form. Unlike bare rock, soils retain water long after a rain and hold on to high concentrations of carbon dioxide elevated by decomposition and root respiration. Unlike bare rock, a soil provides more surface area for chemical weathering within it, because it incorporates rock fragments—resisting *physical* erosion—as the bedrock breaks apart.

For the moment, let us assume that Schwartzman and Volk are correct. Let us grant the possibility that, absent the biotic enhancement of weathering, Earth today would be a hothouse, "favorable, tolerable, comfortable, fit" only for thermophilic (heat-loving) organisms. Bacterial microbes, such as those from the domain Archaea that thrive in Yellowstone's hot springs, would surely rule the world. They would have no competitors. The complex proteins that build the bodies of later and larger forms of life would be ripped to pieces by the jostling of molecules at such high temperatures. Because thermophilic life is not *our* kind of life, we find the idea of an exclusively microbial planet decidedly less appealing than today's riotous display of frogs and forests and fairy shrimp. But are we perhaps self-centric when we judge an entirely microbial world to be missing something? Wouldn't a hothouse Earth that supported thermophilic bacteria in profusion still be "favorable" to life? Yes in the short term, but no in the long term. An Earth that resisted the effects of a brightening sun would stay habitable to some form of life for a longer time than would a biotically passive Earth. Thermophiles are indeed the ultimate biological destiny of this planet. But thanks to the biotic enhancement of weathering, that destiny is billions of years away.[10]

Even accepting biotic enhancement of weathering as crucial for the longevity of the biosphere (not to mention the appearance of anything that can slither, sing, or stand tall in the sunlight), a question remains. Is the cooling role played by life evidence of a *self-regulating* biosphere?

There are only three possible routes for the biotic effect to have taken. Life might have had no effect on greenhouse conditions; life might have exacerbated the brightening of the sun; or life might have offset the solar effect. Is it so surprising that the latter turned out to be the case? Those who regard this turn of events as rather unsurprising will not be driven to call upon Gaia for an explanation. "Just lucky, I guess," will be answer enough.

Researchers somewhat friendly to (or even inspired by) Lovelock's Gaia thesis began to make discoveries of life's powerful influence on the chemistry and climate of Earth. Each discovery—that plankton in the sea emit molecules that foster cloud formation,[11] that the spread of bogs might trigger ice ages[12]—was interesting in itself. But those disinclined toward Lovelock's idea from the outset would probably be drawn no closer to a gaian worldview. "Just lucky, I guess," would remain the refrain. To them, Gaia would still be a false (if scientifically fruitful) hypothesis.

Gaia's begetter thus came to realize that until he could demonstrate the general *mechanism* of self-regulation, the *phenomenon* of an allegedly self-regulating biosphere would largely be ignored. What coccolithophores do to sulphur, and what happens to global albedo when sphagnum moss kills trees, would remain curiosities—tidbits of biogeochemical knowledge drawn from the vast realm of the way things are. Lovelock thus had to find some way to make the idea of Gaia seem less fanciful, less magical. He had to find some way to show that self-regulation at the global scale is, in fact, rather routine. Otherwise, his hypothesis would remain a lovely metaphor, attracting perhaps more spiritual than scientific attention.

"I knew there was little point in gathering more evidence about the now-obvious [to him] capacity of Earth to regulate its climate and composition," Lovelock recalls. "Mere evidence by itself could not be expected to convince mainstream scientists that Earth was regulated by life. Scientists usually want to know how it works; they want a mechanism. What was needed was a gaian model. I wrestled with the problem of reducing the complexity of life and its environment to a simple scheme that could enlighten without distorting. Daisyworld was the answer."[13]

Daisyworld

Daisyworld is an imaginary planet seeded with just one form of life: daisies. The daisies come in two varieties: black and white. In the Daisyworld model, we are unconcerned about chemistry or the water cycle; we are interested strictly in the effect of life on global temperature. Daisyworld orbits a star much like our sun, and so rising temperatures would challenge the longevity of the Daisyworld biosphere. As the star ages, at some point the planet becomes warm enough for the daisy seeds to germinate. Early on, the black daisies have an advantage, because their darkness absorbs the starlight and turns this energy into heat. They are thus more likely than the white daisies to survive and reproduce, soon blanketing the world. Later, the white daisies are favored. Their talent for reflecting the intensifying starlight keeps them comfortable. Crucially, both forms of daisy do more than just heat or cool themselves. They change the temperature of their surroundings; en masse, they tweak global temperature up or down. Thus the planet at first is biotically heated by black pigment spread across the surface; an eon or so later it is cooled by an absence of pigment. The key here is that natural selection alone, operating at the scale of individual organisms, would accomplish a climate boost in the beginning and a moderation toward the end.

Importantly, there is no sensor and decision maker at the global scale. There is no foresight and planning. There is no worldwide altruism. There is no planetary patrol. Each form of life ensures the well-being of the biosphere by doing nothing more generous than growing and propagating whenever it gets the chance. Self-regulation by the Daisyworld biosphere is therefore far from magical. It is an emergent property of a world governed by routine natural selection.

Lovelock and coauthor Andrew Watson presented the results of the Daisyworld model in a 1983 paper subtitled "The Parable of Daisyworld."[14] Why not "A Model for Climate Stability Driven by Natural Selection"? One reason is obvious. It is the same reason why Lovelock years before was delighted to come upon a good four-letter word, Gaia, for a difficult concept. Choosing a flower to power planetary thermostasis, moreover, was a brilliant way to stress that regulatory strength may be the gift of even the most delicate creatures. But why not Tulipworld?

As part of this book project, I asked many of the people I talked with whether they could remember a particular experience at an early age that called up a profound, even mystical, bond with nature. (I was curious about whether biophilia must be nurtured in childhood in order to blossom in an adult.) Jim Lovelock responded that he had those sorts of bonding experiences "on numerous occasions," and he unhesitatingly added that he regards them as religious. "Pre–World War II, I used to go on long walks—up to forty-two miles at a time—every weekend. I vividly remember stepping out one morning in May and walking through a field by a little churchyard. The field was a massive expanse of buttercups and daisies. The whole thing was so glorious. The world, everything, seemed so heavenly."

Whether one finds the Daisyworld model a convincing demonstration of planetary self-regulation depends perhaps on whether one has a problem with the way the whole thing starts. If the artificial jump-start of complementary seed types (evolved out of nowhere) seems a trivial concern, then welcome to the gaian worldview. If not, then a skeptic one must remain. Again, are we "just lucky, I guess" that earthly evolution has given birth to the chemical or thermal equivalents of black or white daisies whenever required—or perhaps in magnificent profusion early on, and repeatedly, so that the seeds are sowed, so to speak, long before the need arises? This is a question that the data alone do not answer.

Overall, some scientists find the Daisyworld model wildly exciting—not only lending insights about the ways of the biosphere but also suggesting new techniques for addressing stubborn problems in the older sciences of ecosystem ecology and population biology.[15] Other scientists, however, are unmoved.[16] Choose your expert, therefore, and choose your worldview. At the moment, the issue remains deliciously unsettled.

In sorting through the arguments, we might well heed a reminder given by the biologist Stanley Salthe. "Self-organization seems miraculous only because we have a habit of not taking it as a primitive fact."[17] Our underlying expectations are Newtonian to the core. Inactivity is the norm in a universe governed by the laws of inertia and entropy. Billiard balls do not move themselves. Grudgingly we admit that life may arise in this universe—but only after umpteen configurations of simple molecules have bumped into one another by happenstance. Even more grudgingly might we entertain the notion of Gaia, self-organization at the global scale. The rise of what has come to be called the science of complexity, however, has put a few cracks in this do-nothing worldview. Biologist Stuart Kauffman, in particular, is pressing for a self-organizing and complexifying universe to be regarded as standard.[18] But the jury is still out.

The problem posed by a brightening sun is thus one challenge to earthly life that seems to call for at least a modicum of geochemical good luck, if not outright gaian management along the lines of Daisyworld, or, more specifically, the biotic enhancement of weathering described by Schwartzman and Volk. Are there any other challenges to the long-term persistence of life?

Shiva, Vulcan, and Life Itself

As with the brightening sun, a second challenge is also external to Earth. This is the risk of a comet or meteor strike that wreaks havoc with the planet's chemical and thermal regimes. An intruder from space just ten kilometers in diameter is now held responsible for the sudden demise of the dinosaurs and many creatures of the sea, including the beautifully shelled mollusks called ammonites. An impactor far smaller would put humans back to the Stone Age. The darkness caused by particles of ejecta blocking the sun would cause crops to fail everywhere and temperatures to plummet. Water supplies might be ruined by acid rain. Social chaos would accompany mass starvation.

Geologist Michael Rampino is at the forefront of researchers looking for evidence that such aliens from space triggered perhaps all of the Big Five episodes of mass extinction on Earth and numerous lesser catastrophes, as well. In playful counterpoint to Lovelock's Gaia, Rampino calls upon the Hindu god of destruction and rebirth: Shiva. "The theories [Gaia and Shiva] seem to occupy two ends of the spectrum of our view of the natural environment. In one the environment is cared for and preserved by the biosphere, which is capable of responding to relatively slow geological and astrophysical changes. In the other, the biosphere is overwhelmed by sudden catastrophes from space that could, theoretically, snuff out life on Earth entirely, and perhaps has in other places in the universe."[19]

The sudden and random challenge of collision with a cosmic object thus contrasts with the ever-present, if exceedingly slow, challenge of a brightening sun. Both are trials initiated by forces well beyond

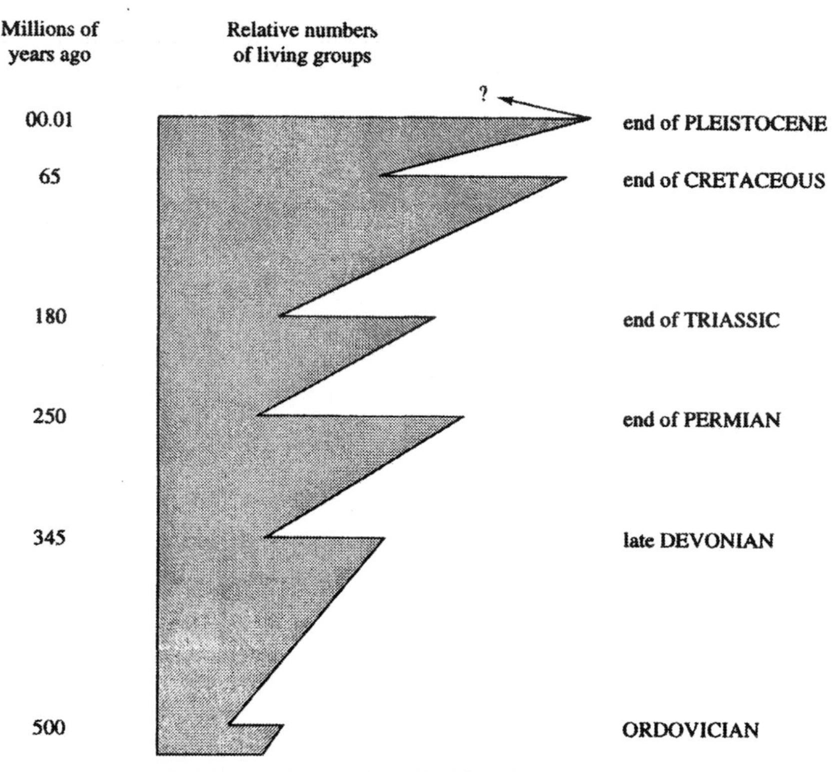

THE SIX MAJOR MASS EXTINCTIONS. Adapted from Richard B. Primack, 1993, *Essentials of Conservation Biology* (Sunderland, Mass.: Sinauer), p. 78.

the confines of Earth. But what about internal challenges? Does Earth itself ever present life with a problem of global scale?

Gaia, Shiva—what about Vulcan or Hades? The geological record tells us of catastrophic episodes of volcanism. A form of volcanism that hominids have never witnessed is perhaps the most capable of precipitating biotic disaster on a regional and even global scale. Rather than coming up through a single volcano, lava gushes out onto a continent through tens or hundreds of miles of cracks. Inconceivable volumes of molten rock are moved from the realm of the underworld to the realm of Gaia over a very short stretch of time, geologically speaking. The result is a vast *flood basalt* rolled across the land and a catastrophic volume of carbon dioxide, noxious gases, and particles belched into the sky.

Residents of Washington state who live east of the Cascade Mountains encounter all sorts of reminders of a flood basalt 16 million years old, swept clean and sculpted in many spots by a catastrophic release of glacier-impounded water during the waning of the last ice age. Far bigger is the Deccan flood basalt of India. This extrusion is dated to right around the end-Cretaceous dinosaur extinction, 65 million years ago; the crackup of that section of crust may have been related to a meteor impact half a world away, along the Yucatán Peninsula of Mexico.[20] The biggest flood basalt of all time lurks in Siberia. This Siberian basalt happens to coincide with the time of the greatest mass extinction ever—the one that put an end to the Paleozoic era 250 million years ago.

If the biosphere has indeed been subjected to episodes of what Diane Ackerman calls "cataclysms galore,"[21] then how did life—particularly more complex, and hence more delicate, forms of life—make it through? Are the recoveries following meteor- or volcanic-caused mass extinctions examples of gaian self-organization? If so, *how* does self-organized recovery of the biosphere actually happen? Or are these recoveries tributes to the sheer power of life at the scale of lineages to adapt and to fill vacant ecological niches? Even with all the data one might hope for, how could a scientist tell the difference between a gaian-caused recovery and a strictly darwinian one?

Power can run amok, too. Thus a fourth and final challenge to the persistence of life on Earth is life itself. This is the opportunistic creep of evolution. Innovations that dazzle with their ingenuity and the survival edge they give their progenitor might, nevertheless, pose real problems for an ecosystem and perhaps even the biosphere as a whole. Every significant novelty must have its counter. From the standpoint of chemostasis, there are two paths to perdition. First, somebody gets creative and starts dumping a new kind of toxin that nobody yet can detoxify. Second, somebody hoards an essential nutrient by taking it with them beyond the grave—either by constructing an organic molecule that cannot be broken down or by triggering a physical process that sequesters the nutrient in a chemical form or a physical location that life cannot access.

Both poisoning and hoarding have, in fact, happened on a global scale in biospheric history. The biotic production of molecular oxygen is the prime example of a toxic catastrophe. The primordial anaerobic (oxygen-free) atmosphere was utterly transformed by a lineage of microbes that discovered how to use sunlight to power a new form of photosynthesis, by which the hydrogen essential for building carbohydrates was mined from water, rather than from the much scarcer supply of hydrogen gas produced by volcanoes. This oxygen catastrophe is widely regarded as the greatest pollution event of all time. Anaerobic forms of life that once roamed the surface of Earth were thenceforth confined to subsurface sediments, only much later profiting from the paradisiacal homes made available to them in the guts of termites, cattle, and other herbivores that can't digest fibrous foods without them.

Life did, of course, pull through the oxygen crisis, turning a big problem into a big opportunity. Aerobic forms of life originated and then evolved in splendid diversity. An ozone shield was born, allowing intertidal and land life to be a lot more casual about hanging out in the sun; photosynthesizers could now capture the desirable wavelengths without life-threatening exposure to the ultraviolet end of the spectrum. Also, because oxygen is an immensely powerful way to burn food calories for energy, complex nervous systems became a realized possibility.

The toxic challenge of oxygen release was thus ebulliently overcome. The record may be less admirable, however, for two episodes of hoarding. First, when the animal kingdom evolved digestive tracts some half-billion years ago, they began to excrete food wastes in the form of fecal pellets. From then on, the open ocean ecosystems may have lost a lot of nutrients to gravity. Drifting down into the darkness of the depths, the nutrients borne in feces would no longer be readily available to any photosynthesizer, thus introducing a new challenge for life.[22] A second possible example of an evolutionary misdeed that resulted in hoarding happened in the plant kingdom, on land. This was the invention of lignin—wood. Lignin was invented in the late Paleozoic and made the tall, tree-form possible. But lignin is also notoriously difficult to decompose. Certain kinds of fungi do a decent job of it today, particularly in wet climates, but there may have been a serious pile-up of wood in the early history of trees, before the fungal kingdom devised a way to dine on it. Do the world's vast Carboniferous coal deposits testify to such a hoarding event?[23]

So, life or Gaia may be less than perfect. So what? Consider the achievements! Hoardings and poisonings fade into oblivion in comparison to three and half billion years of pulsing life. The biosphere may have its problems, but on the whole it works and works with gusto.

In your lifetime, the full complement of carbon dioxide in the atmosphere will be taken up by photosynthesizers and replenished by respirers perhaps ten times. Because fifty times more carbon dioxide is dissolved in the sea, a single biotically driven flush that ends today would have got its start around the time Newton was born. Even this rate suggests a whirlwind of activity. Just the tiniest imbalance would upset the chemostasis and thermostasis of Earth in the blink of an eye, geologically speaking—and that is what we humans seem to be doing right now.

Call it good luck, the power of life, or the genius of Gaia. The persistence of life on this pocked and parrying Earth is cause for astonishment.

A Developing Biosphere

The science of the whole Earth as a living system is still in its infancy. Today's gaian researchers can therefore experience the same thrill of discovery that zoologists and botanists enjoyed during the heyday of natural history nearly two hundred years ago—when Alfred Russel Wallace and Charles Darwin were collecting shiploads of new species in the Southern Hemisphere. A better analogy might be to the early days of physiology, more than three hundred years ago, when William Harvey and others were discovering how the human body works. Will our own times be remembered as the dawn of *geo*physiology?

The geophysiologists pioneering the science today are, necessarily, amateurs to some extent. Those educated in biology have had to scrabble for an understanding of geochemistry. Those trained in geochemistry have had to scrabble for an understanding of physical oceanography or atmospheric science. One solution to the problem of neuronal overload is, of course, collaboration.

Collaboration made it possible for me, a thoroughly amateur amateur, to actually make a contribution to gaian science. In 1988 I had an idea and tried it out on a scientist I had met earlier that year at the Gaia conference sponsored by the American Geophysical Union in San Diego. The idea seemed to have merit, and my confidant had an office full of essential references, so Tyler Volk and I teamed up for what we eventually titled "Open Systems Living in a Closed Biosphere: A New Paradox for the Gaia Debate."[24] It occurred to us that the faint young sun paradox was not the only puzzle that called for either a gaian or a "just lucky, I guess" explanation. How is it, we asked, that in three and a half billion years of biological evolution, the finite stores of Earth materials have never been depleted or poisoned by life?

All organisms ingest nutrients and excrete wastes, but the biosphere is the only living system that must completely recycle everything. Oh yes, a few errant baubles from space intrude from time to time, matter (such as carbon) can be put in deep storage, and volcanic outgassing must be dealt with, but on the whole the biosphere must abide by the principle that what goes around, comes around. There is no place to shop for fresh supplies if nutrients are squandered. There is no place to dump toxins safely, out of sight and out of mind. Thus the paradox: All living systems are wide open to matter exchange, but the biosphere is a living system that is very nearly closed.

Scientists working with the U.S. and Russian space programs have for years tried to construct simple analogues of the biosphere, sealed glass containers powered just by sunlight (and sometimes artificial light). What they have found is that a recipe of ingredients has yet to be concocted that can support a vertebrate for any length of time (recall the problems encountered at Biosphere 2). Shrimp might live to a ripe old age in a baby biosphere, but all hell breaks loose if they make the mistake of reproducing. Microbes, however, may hold on for perpetuity.

Building a baby biosphere by using only the organisms our eyes cannot see may, in fact, be more like improvising a soup than composing a soufflé—just about anything will work. Begin with a glass vessel; any jar or bottle will do. Add water, leaving room for a miniature atmosphere at the top. Toss in a little garden-variety fertilizer and a good dollop of "starter"—mud from your favorite marsh. Then seal the contact with candle wax and set the container in indirect sunlight (you don't want to cook the contents).[25] Chances are, the potion will bubble and thrive, outlasting the human patience required to keep an eye on it.

Bacteria are more than just adept at surviving tough conditions. Bacteria run the show. More than any other life form, bacteria have the right to claim, "We are the world." Metazoans such as nematode worms, scorpions, rattlesnakes, and people are lovely but, alas, expendable. Our greatest contribution is providing sturdy housing for bacteria—the anaerobic palaces of our guts and the cozy cellular enclaves that are home to remnant bacteria of an ancient lineage, the mitochondria.

Consider: Nitrogen is a vital ingredient in each and every protein and DNA molecule made by life. Yet no plant or animal or fungus or protist can "fix" molecular nitrogen drawn from the atmosphere into a form usable by life. None is capable of sundering the tough triple bond of the dinitrogen (N_2) molecule and linking those atoms with hydrogen atoms to form ammonia (NH_3) or ammonium ions (NH_4^+). Several hundred atmospheres and 500°C are the extreme conditions we industrial humans must provide for the reaction to take place in our fertilizer-manufacturing plants. Certain strains of bacteria can do it, however. They do it on their own, or in special housing constructed just to their liking by plants and algae, who are eager hosts.

Getting nitrogen out of the atmosphere and into living tissue is one task; putting it back is another. There, too, bacteria excel. Depicted here is a schematic of the nitrogen cycle. I offer it both for the scientific content and as a spiritual mandala. Contemplate it, and thereby contemplate the marvel of Gaia. Perhaps before sinking into a deep meditation, begin by noting that most of the cycling activity is achieved by life and that life—the ultimate opportunist—simply patches in to the fragments of cycles that are the free gifts of geochemistry. Now follow what happens when you remove all the biologically driven arrows: The nitrogen gets stuck in the form of nitrate (NO_3^-) and nitrite (NO_2^-), which are rained into the oceans.

The symbol of life at the biospheric scale is surely the cycle. Ele-

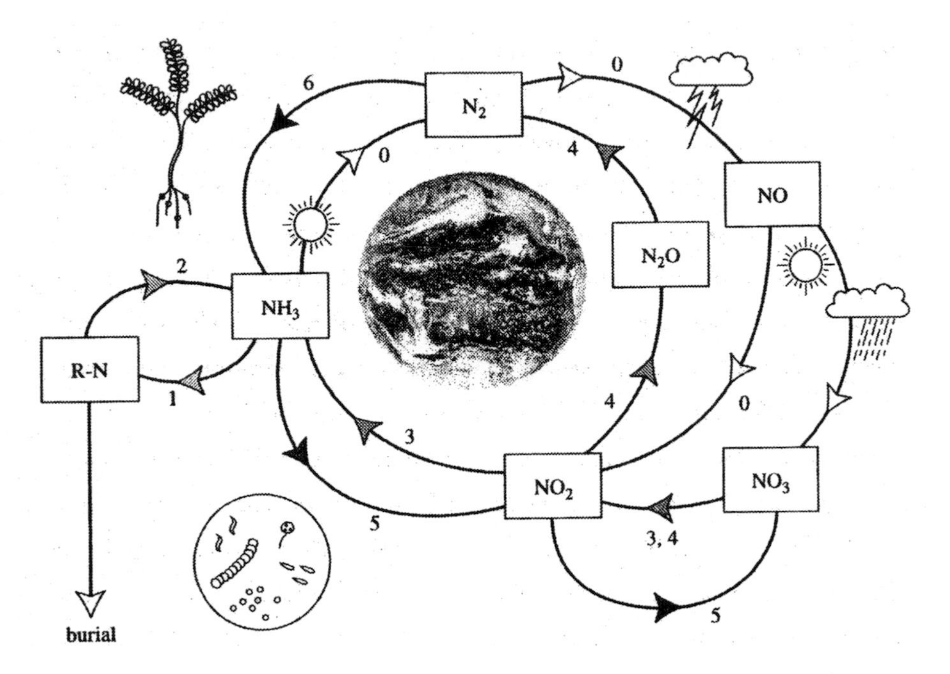

A GAIAN MANDALA: THE NITROGEN CYCLE. The sequence of developmental stages begins with (0) prebiotic lightning, solar-induced changes, and raining out of particulates. The remaining stages are all evolutionary innovations by life: (1) ammonia consumption, (2) ammoniafication, (3) nitrate assimilation, (4) denitrification, (5) nitrification, (6) nitrogen fixation.

ments crucial to life begin in one compound and then are worked by a combination of living and nonliving processes through a series of other compounds, eventually returning to the first. Life at the biospheric level circumscribes a family of cycles, notably, those of nitrogen (N), carbon (C), oxygen (O), hydrogen (H), phosphorus (P), and sulfur (S)—all of which are essential for life. Visualizing these interlocked rings, glistening conduits of molecules in motion, may not be quite so sublime an exercise as visualizing the Hindu image of Indra's net of jewels, but to my mind it is a close second.

The closed biosphere paradox is a puzzle to be approached scientifically for an answer. It is also a koan, offering the quester who follows the way of science a chance to experience the wonder of life at the scale of a whole planet. Don't expect the scientists themselves to point out that route, however. That is not their job. Those who wander too far down the extensionist path can count on the criticism of peers. It happened to Jim Lovelock in the wake of his first book. But it will surely not happen to Lynn Margulis. I spent a good portion of an hour trying to pry such extensions out of her, but she stood her ground. I was not surprised. Some of her most memorable writings aim at presenting Gaia as the least miraculous of all scientific explanations of planet-scale phenomena. "One invokes either physiological science, magic, or God," she reiterated, "to explain Earth's wildly improbable, combustible, thoroughly drenched troposphere. The Gaia hypothesis, in acknowledging this atmospheric disparity and seeking its causes, has opted for physiology over miracles."[26]

Margulis has been a vital player in the formulation of Gaia theory. She credits Jim Lovelock exclusively with the Gaia idea (despite three early, much-cited papers jointly written with him,[27] and despite his ever-willingness to share the acclaim). Her contribution to the gaian research program has been twofold. First, she extends to the planetary level an understanding of the permanent associations of living beings through symbiosis and the synergetic consequences that result when individuals from different species intimately connect their livelihoods. She credits one of her former students, Gregory Hinkle (now a professor at the University of Massachusetts, Dartmouth), with the metaphor "Gaia is symbiosis as seen from space."[28] Second, she speaks for the power of the omnipresent yet often-overlooked microbial realm. Whether in mat communities or sequestered as mitochondria and plastids in the cells of larger organisms, it is the bacteria—more than any other kingdom of life—that keep the gaian cycles going.

"For more than a billion years," Margulis reminds us, "the only life on this planet was bacterial. Today, even through a powerful microscope, all bacteria look very much alike: boring or menacing from a human-centered vantage point. However, bacteria are the source of growth and reproduction, photosynthesis, atmospheric methane production—really, they invented all the interesting features of life except perhaps speaking Spanish. Bacteria and their symbiotic derivatives, protoctists, are still with us in large diversity and numbers. Microbes still rule Earth."[29]

From a gaian perspective, therefore, the most exciting events in the evolutionary epic took place long before the first trilobite graced the Cambrian seas. Microbial communities may, in fact, hold the answer to how a coherent biosphere could have evolved. Ever since Lovelock launched his idea, critics have pointed out that "fitness" at the planetary scale cannot be attributed to evolutionary processes because there is only one biosphere.[30] Natural selection cannot take place in a population of one. This is because darwinian selection requires a set of replicators in which culling—survival of the fittest—can take place. Life doesn't know in advance whether a particular inherited change will prove beneficial. There is no foresight and no planning in evolution. Blind trial and error, made possible by the fecundity of excess, is at the root of evolutionary creativity. But there is only one Earth. There is only one biosphere. One lethal mistake from a planetary perspective, and life on Earth ends. There is thus no room for trial and error in a population of one. This is why microbial mats may have been exceedingly important on the early Earth.

Billions of years ago, microbial mats probably invented at a very small scale the nutrient cycles essential for life. Consider that microbial mats today must cope with the desiccation that comes at low tide. They must screen out ultraviolet light, which would have been a far greater problem billions of years ago when Earth lacked an ozone shield. And they must avoid losing precious volatiles to the atmosphere or ocean. For all these reasons, mat communities, then as now, covered themselves with mucopolysaccharides—slime—as sealants and light filters. Microbial mats are thus partially closed to gas and liquid exchange with the outside world. Long ago, member bacteria overwhelmed in waste who discovered ways to convert the waste into anything useful for themselves would have grown prodigiously—at first. Such innovative microbes would make their own peculiar waste, thus exerting selection pressure on yet another kind of bacterium to find ways to remove the waste for personal gain or sheer survival. One microbe's waste thereby became another's food. Diversity, forced by the exigencies of existence and growth, would have blossomed.

Mini Gaias, in the form of microbial mat communities, thus pioneered new metabolic pathways, which were judged by natural selection at the scale of the individual microbe and the community as a whole. Eventually the external atmosphere came to resemble the internal atmospheres of the most successful, and thus most prolific, microbial mats and other integrated communities of bacteria. Gaia would emerge naturally by way of the stringent demands of natural selection—operating at the microbial rather than planetary scale.

Nobody speaks more favorably, even lovingly, of microbial mats than does Lynn Margulis. And she has been a mentor for me since, as American editor of the international journal *Biosystems*, she tinkered with Tyler's and my closed biosphere paper and accepted it for publication. At the time, she suggested a second name for the paradox: the Vernadsky paradox.

Vernadsky who?

Thus began my education in the life and work of the man one biographer has called "the most prominent Soviet natural scientist of the early Twentieth Century."[31] Because of the language barrier and the intellectual chasm caused by the Cold War, Margulis and Lovelock had been unaware of the legacy of Vladimir Vernadsky when they formulated the Gaia hypothesis. Years later they discovered they had traversed much of the same ground but that they had also pressed beyond, into terra incognita.

Vernadsky's pivotal insight was that "life is the greatest of all geological forces." Several generations of Russian scientists have been schooled in the tradition pioneered by this founder of biogeochemistry. Their discoveries of how living organisms work alchemical wonders on Earth's crust make science fiction seem rather tame.[32] Vernadsky shared with Margulis the microbial perspective: The geochemical effect of physiological activity is inversely proportional to body size, he observed. So what good are animals? The "cosmic role" of metazoans, from burrowing worms to bipedal humans, is geophysical rather than geochemical. Our job is biotransport.

Microbes chemically alter the crust, but only we can move it in directions that gravity, wind, and water refuse to carry it. A shoal of chinook salmon, working through rapids a thousand miles up the Yukon River, is life as a geological force. Millions of mayflies emerging en masse from streams, carrying nutrients back onto the land; Kentucky skies darkened for three days by a migrating flock of passenger pigeons; one part of Africa denuded by locusts who then carry the bounty to another; juvenile sea turtles powering their way out of the calm cradle of the Sargasso Sea—all these are exquisite contributions of Kingdom Animalia to the churning of the biosphere. Margulis playfully presents this Vernadskian perspective in her depiction of albatrosses as "meter-wide phosphorus-nitrogen-sulphur-rich locomotory volatiles in the atmosphere."[33]

What do we big-brained two-leggeds do for the biosphere? Like our chimpanzee kin, we started out as pretty good dispersers of fruit seeds. Excrement was our precious gift to Gaia. Now we have become mighty harvesters of stored phosphates, liberators of troves of minerals (often originally concentrated by microbes), and diverters of rivers onto parched landscapes. Vernadsky was a big booster of the biospheric benefits of industrial civilizations. He lived in the era and in one of the empires that saw no downside in the path of progress that stretched out ahead. Today, however, his unquestioning faith in the goodness of biospheric manipulation seems as chilling as it is naive. The more we learn, the more we sense the folly of such hubris. Nevertheless, Vernadsky's insight that life as a whole is a geological force transcends the limits of his time.

How is Jim Lovelock's contribution distinctive from that of Vernadsky? Where does Lovelock stand?

Perhaps with Newton, on the shoulders of giants. Charles Darwin saw life as shaped by its environment. Vladimir Vernadsky saw the environment as shaped by life. Jim Lovelock connected the two arrows of cause into a cycle of feedbacks. Out of that cycle emerges gaian self-regulation. This is Lovelock's imaginative leap—though the jury is still out as to exactly how much of his vision, too, will stand the test of time.

Lynn Margulis notes that whereas Vernadsky perceived life as matter in motion, Lovelock enlivened matter.[34] From a gaian perspective even the atmosphere comes alive, "like a cat's fur, a bird's feathers, or the paper of a wasp's nest." The two perspectives differ in other ways, as well. How planetary change should be regarded is perhaps the prime example. Lovelock is in awe of Gaia's remarkable stability, the biosphere's resistance to change—particularly in light of the sun's increasing luminosity. Vernadsky, in contrast, extolled the long-term trends he perceived. Opposed to the Lovelockian cycle of stability is thus the Vernadskian arrow of change, of biospheric progress.

A rough sketch of the Vernadskian viewpoint of planetary development is that, with time, the biosphere has been getting bigger, faster, and smarter. *Bigger,* in that more and more types of chemical elements and more and more volumes are being swept into the stream of life, including the vast life-created storehouse below, which Vernadsky poetically called the realm of "bygone biospheres."[35] (Because of our modern understanding of plate tectonics, we know that materials buried in the realm of bygone biospheres are only in temporary storage.) *Faster,* in that life finds ways to capture more and more of the available sunlight, further energizing the frenzy of movement. *Smarter,* in that Vernadsky had faith in what he (and Pierre Teilhard de Chardin) came to call the noosphere—the sphere of mind and intelligence that, culminating in humankind, was now powerfully affecting the other, much older, spheres (atmo, hydro, litho, bio).

Land life excels in the "bigger" mode. Owing in large part to the need for land plants to use cellulose and lignin as support in the flimsy fluid of the atmosphere, and owing to the arms race that forest trees engage in to overtop their neighbors and thus capture more sunlight, the weight of all living material on land is two hundred times greater than that found in the sea. Moreover, if the relative proportions of producers and consumers in these blocks of biomass are compared, the two realms are shockingly different. A "food pyramid" that represents the land biomass (with producers as the foundation and consumers higher up) has a very wide base indeed. In the ocean, however, the food pyramid is more an obelisk. The producer:consumer biomass ratio on land is 600:1; in the ocean it is only 4:1. How is this possible? *Faster* is the answer. The average alga is born, reproduces, and is swallowed up by an animal in just a few hours or days. Because these single-cell organisms reproduce so quickly, food is ever available for life forms perched higher on the obelisk. What this means is that *productivity* in the ocean is not that much less than productivity on land. Each year about 120 gigatons of carbon dioxide are photosynthetically fixed into carbohydrates on land (and released by respirers). This compares with 60 gigatons of carbon dioxide drawn down annually by sea life. Land and sea thus pair with the bigger and faster biospheric principles of Vernadsky. But the key question is whether, throughout the course of evolution, the biosphere as a whole has, in fact, been getting bigger and faster.

"I'm not convinced, actually, that the biomass has increased through time," says David Schwartzman. "It may have maxed out early on. We don't know what the biomass of the ocean was several billion years ago. The geochemical record is ambiguous in being able to answer that question now." David Schwartzman is a geophysiologist who credits a strong Vernadskian heritage. It is Vernadsky's general insight of life as a geological force and the developmental perspective that attract Schwartzman—not the specifics of any of Vernadsky's claims about trends. "The sense I got of Vernadsky was that the whole process was an arrow; it was going somewhere. You see, it was a directional thing. Life had a progressively increasing role in affecting the crust and the atmosphere."

"That's a nice, inoffensive way to use the word *progress*," I respond. "Life is playing an increasingly progressive role in affecting Earth—not that there was progress per se." David Schwartzman and I, friends through Tyler Volk (my coauthor in the closed biosphere paper), are sitting on a grassy knoll in a park in Washington, D.C., that overlooks a soccer field where his son is practicing. Trained as a geochemist, David now regards himself as a geophysiologist, as well. He is a professor in the Biology Department at Howard University.

"I don't buy into the so-called laws that Vernadsky proposed," he continues. "But I do see a developmental trend."

"And it's not a trend that Vernadsky identified," I add.

"No, he couldn't have detected it. He didn't have the data."

What is the developmental trend that Schwartzman perceives? *Cooler.*

Schwartzman agrees with many geochemists in positing a hot early surface temperature of Earth, produced by a dense atmosphere of greenhouse gases four billion years ago. He parts with mainstream views in his conclusion that life itself played a major role in cooling the planet, by way of biotic enhancement of weathering. Cooling, moreover, apparently occurred over several billion years, with evolutionary innovations playing a key role. The most pronounced drop in temperature happened between 1.0 and 1.5 billion years ago.[36] Remember that, absent some drawdown of the initial complement of greenhouse gases to match the brightening sun, Earth would instead have been getting warmer through time. Schwartzman believes that life has done far more than counter the sun toward some cozy, homeostatic end. It has done that job and then some. I ask him for a recap of his story of discovery.

"Let me start with what isn't new. What isn't new is that people have, in fact, been getting evidence for a hot early Earth that lasted well into the Precambrian. This derives from oxygen paleotemperature work on cherts, which are silica rocks preserved under marine conditions. Cherts retain the signature of the fractionation of oxygen isotopes at the sediment–water interface. And that fractionation gives us an indication of temperature. The chert work was begun in the mid seventies by Paul Knauth and Sam Epstein."

These chert interpretations are, however, considered an anomaly and thus are ignored by most geochemists with expertise in paleoclimates. The widespread, mainstream assumption is that if in fact Earth at the origin of life was hot, then it rapidly cooled. For all intents and purposes, the climate has for a very long time been pretty much as it is today—bracketed on one end by somewhat warmer times (such as during the Cretaceous, when crocodilians thrived in the arctic) to occasional glaciations (which we are still in to some degree, given that Antarctica is under ice). This rather homeostatic view of climate history is supported by what is widely accepted as rock-solid evidence:

glacial tillites dated to about 2.3 billion years ago. This is the so-called Huronian glaciation. David Schwartzman accepted that glacial evidence not long ago. He was thus part of the paleoclimate mainstream.

"What triggered my re-examination of that was a conversation I had in your and Tyler's apartment, with Michael Rampino. This was in November 1992. It turns out that Michael had been mulling over his hypothesis for years, that most of the ancient glaciations—that is, not the Pleistocene, but the ones going back into the Paleozoic and Precambrian—were not glaciations at all. They were impact deposits. That conversation made me re-examine the whole argument about temperature."

As Rampino had explained that evening, until very recently any geologist coming upon a jumbled deposit of boulders and cobbles was likely to call it a glacial tillite. Tillites are the mish-mash of debris that melt out of a stagnant glacier at the end of an ice age. Because the Pleistocene ended just yesterday, tillites are well-preserved and hence well-studied geologic formations. In contrast, how many geologists—especially in decades past—would even consider "impact ejecta" as a possible classification for those same sorts of deposits? How many have even read about the characteristics of such deposits, much less visited any? "Impact ejecta" was thus not in their lexicon. But with today's better understanding of the Ries impact crater in Germany and the shocking discovery of the much larger Chicxulub impact crater off Mexico's Yucatán peninsula, which correlates with the end-Cretaceous dinosaur extinction, it is time for reinterpretation.[37]

"I remember that night," Schwartzman continues, "specifically, going down the elevator, talking with Michael, and saying, 'Well, this removes a key constraint if it's true.' Back in D.C., I started looking at the other constraints—the gypsum record, in particular. In sedimentary rocks as old as three and a half billion years, gypsum has been inferred from the geometry of replaced crystals. It turns out there are good arguments that gypsum precipitates metastably at temperatures far above the value that's supposed to be its equilibrium temperature, both in the laboratory and in nature. So, in my mind, the inferred precipitation of gypsum in the early Precambrian does not, in itself, signify a cool Earth. And that night at your place I learned that so-called glacial tillites don't necessarily signify a cool Earth either."

If Schwartzman is correct in his own reinterpretations, then for two-thirds of Earth history, global climate was very warm indeed. Life would have been exclusively thermophilic. Nevertheless, Schwartzman's new view of paleoclimates will probably remain a curiosity among his geochemical colleagues until Rampino's cataclysmic reinterpretation of deposits now classified as tillites is more widely accepted by geologists. Even then, the idea of a long-lingering hot Earth will have a rough go, because it is so brashly contrary to the reigning paradigm in the earth sciences. Biologists, however, may find it immensely attractive.

"Did you tell Maynard Smith about your stuff?" I ask David. John Maynard Smith, one of the most respected theorists in evolutionary biology, was at the Gaia meeting at Oxford University that we both had attended, as Jim Lovelock values his skeptical views.

"Oh yes. Matter of fact, he referred to the evidence for hot conditions in one of his informal remarks. Remember?" David and I have, by now, moved from the soccer field to his home office, so he reaches over to one of the piles of books that cramp his desk and pulls out Maynard Smith's latest.[38] "See here how this idea can solve the biological paradox Maynard Smith identifies." He begins to read from the book, " 'Remnants of prokaryotes are 3,500 million years old. By contrast, the oldest eukaryotic microfossils are not older than 1,500 million years. Astonishingly, the time that was needed to pass from inanimate matter to life is four times shorter than that needed for passing from prokaryotes to eukaryotes. Although the origin of eukaryotes involves really difficult steps—such as the elaboration of protein import from the cytoplasm, following the transfer of genes to the nucleus—it is hard to argue that they are more difficult than to establish a genetic code.' " Schwartzman looks up, closing the book. "Maynard Smith then goes on to say that this is a real paradox."

"So Maynard Smith is asking, 'Why was evolution so slow?' " I continue, "And you have a most attractive answer. Remember a couple years ago, at the philosophy of biology conference in Illinois, George Williams was very interested in your presentation.[39] [Williams is to evolutionary biology in America what Maynard Smith is in England.] If you are right, then this really helps out the biologists."

"Yes. But, you know, I fluctuate between thinking that I've come onto something big that is really important and that what I've done is a bunch of rubbish. The only thing that gives me some confidence is seeing how it's stimulated other people to do stuff, to test certain things. And I think it has."

How exactly does Schwartzman's idea help evolutionary biologists? He published thoughts on this topic in the journal *BioScience* (with coauthors Tyler Volk and paleontologist Mark McMenamin) in 1993.[40] The gist is that each stage in the evolution of life happened *as soon as it could happen*. If the record of life's accomplishments seems a bit slow, it is through no fault of biology. High temperatures are to blame. What is more, life itself was ever working to reduce those temperatures, by enhancing chemical weathering of silicate rocks.

Schwartzman continues, "Cyanobacteria, with an upper-temperature limit for growth today of 72°C, appear in the fossil record when 3.5 billion year old cherts indicate an environmental temperature of about 70°C. For the aerobic, mitochondrial eukaryotes, the temperature limit for growth is about 60°C. Sure enough, their appearance also correlates with the chert record."

"So until greenhouse gas levels were reduced enough to bring the temperatures down to 60°C," I reiterate, "eukaryotes simply could not evolve, try as life might. That does seem to resolve the Maynard Smith paradox."

"I don't know if there is anything in this hypothesis pertaining to the temperatures of later events," Schwartzman cautions. (Mean global temperature today is about 15°C.) "Maybe there is some connection between evolutionary events and temperature in the Phanerozoic [starting about 540 million years ago, with the Cambrian period], but I'm not going to speculate at this point. You still might have had a temperature constraint for certain megascopic life, say for vertebrates, but I don't think it's really been explored. I've seen quoted in the literature that the upper temperature limit for marine invertebrates is 38°C, but that's for present-day populations. The upper temperature limit for metazoa is somewhere around 50°C. What determines the upper limit of metazoa? Why is it 10° lower than for thermophilic fungi, that go up to 60°C? It is thought that these limits are biophysical, biochemical, but exactly what causes them is not yet clear."

"So evolutionary innovations tell us something about the temperature at the time they occurred," I conclude.

"Yes. The thesis is that the surface temperature at emergence of a new stage of complexity corresponds to the upper temperature limit for growth of that group."

"If you're right," I say, "then scientists someday will be building their paleoclimate profiles not on chert or gypsum data or misclassified glacial tillites but on fossil life."

Schwartzman leans forward. "I think the whole history of the biosphere is embedded in the genome, if we can only learn how to interpret it. We see this in the deep-rootedness of the thermophiles. We see this in the anaerobic/aerobic transition. I'm sure there are other transitions in history preserved in the genome, if we can learn to interpret it."

The technical talk continues, and then Schwartzman concludes, "A lot of evidence seems to be converging on supporting this story. There might be some bombshell that blows it all away, but so far I haven't seen it."

"Okay," I say, "Let's assume that twenty years from now Michael's impact idea is standard. Ancient glaciations are just a figment of an outdated imagination. And so let's say that you have convinced the geochemical and biological establishment to come your way. Science watchers like me now have to embrace the notion of a long cooling trend through time. But what do I get out of it spiritually? Why is cooler in some sense better than warmer? Is it because the biosphere gets to keep going a bit longer before the oceans boil away a few billion years from today? I'm sorry, but that is just too far off for me to care a great deal. Or is it because a cooler climate makes our own species possible? That's just a bit too anthropocentric for my tastes."

"How do you feel about polar bears and penguins?" Schwartzman catches my smile and continues, "Seriously, think about what cooler means. It means that some places on Earth get cooler—some a lot cooler—than ever experienced before, and yet some remain just as hot: the warm springs, the hot springs, the hydrothermal vents on the ocean floor. So a cooling trend means that you get an increase in the diversity of habitats through time. Remember that above about 30°C globally—at least this is Marty Hoffert's view—the planet would have been isothermal. Apparently, at that temperature storms would have been so intense that they transferred the excess heat near the equator all the way to the poles. So a deep cooling introduces even another factor adding to habitat diversity—a factor that wasn't here for most of Earth history: a big latitudinal temperature gradient."

He continues, "Now, the increase in habitat diversity works not just for temperature. Think about oxygen history. At one time there was no free oxygen anywhere; now there's mostly aerobic habitats. But there are still anaerobic pockets all over the place—down in marshes, on the sea floor, in your gut."

"E. O. Wilson says the evolutionary trend is an increase in diversity of life; you say an increase in diversity of *habitats*. Seems to me your version of the story is more fundamental. It goes to the wellspring out of which comes an increase in the diversity of life. It's a more geophysiological, rather than just biological, view of it all. I like that."

It is approaching midnight. I suggest we ease out of this discussion by revisiting the idea of the arrow of biospheric evolution, which is the way we began. "When you use the word *evolution*, as in evolution of the biosphere, you imply an arrow that has a trajectory to it, whereas a lot of people in evolutionary biology are convinced that change is a random walk—a little this way, a little that way. But you say there is a trend."

Schwartzman responds, "Well, at least it appears on Earth. In contrast to evolutionary biologists who focus on the organism, I suspect that some of the main patterns in biospheric evolution may actually be quasi-deterministic. If you run the whole thing again, it may be largely repeated in general patterns—not just the cooling but also the building up of eukaryotes out of prokaryotes, for example."

"Even more basic," I add, "would you agree with me that you're going to have an anaerobe-to-aerobe trajectory because at some point there's going to be a hydrogen shortage, and wherever you are, life is going to want to crack water to get at the hydrogen. And so up goes the oxygen. You're going to have a nitrogen cycle, too, that has to fill out piece by piece through time."

"Sure," he replies. "The metabolism of the biota is going to affect the metabolism of the biosphere. And that's what we call Gaia—the metabolism of the biosphere.

"Remember," Schwartzman continues, "Vernadsky didn't have a concept of self-regulation. That's the real originality of Lovelock's approach. What also was attractive to me about Lovelock's work is that it challenges the paradigm of evolutionary biology very strongly: the paradigm that life simply adapts to its environment. And here we have a theory that the global environment has been profoundly affected by life's activities, and that in turn feeds back on the system. That's dialectical. There's the feedback, and it's not a feedback that's static, that goes in a circle. It's a feedback that evolves like a spiral."

The idea strikes a chord with me. I remember having come across the image of the spiral as symbol of biospheric evolution while reading a book by one of Vernadsky's students.[41] I now understand why the spiral is such a powerful image. The circle represents the regulatory triumph of homeostasis. It also calls forth the spinning of the matter cycles. The spiral view, however, blends the circle of stability, of sameness, with the arrow of evolution, a developing biosphere. It is the synthesis of the two that gives us an even more beautiful symbol of what Earth history is all about.

I have thus come to regard the developmental view of the biosphere as even more alluring than the homeostatic view. Homeostasis is still there, of course, in the millennium-to-millennium maintenance of Earth's atmospheric mix and in the ceaseless flow of molecules through the biospheric circulatory system, but now I can meld Gaia theory with my passion for the evolutionary epic. And that's just great. A geophysiological view of evolution, moreover, brings out the deep patterns—relegating the god of contingency to a subordinate role. On top of that, I get to add the spice of biodiversity in a gaian context: Schwartzman's notion of a trend of increasing habitat diversity through time. All the while I can feel cozy and welcome here because, as Gary Snyder has put it, "we are all indigenous to this planet."[42]

A Conversation with the F_1 and F_2 Generations

David Schwartzman is one of a rather intimate group of scientists that I have come to call *the F_1 generation of gaian researchers*. F_1 is a term drawn from genetics. It signifies the first (filial) generation born to a parent couple of peas, fruit flies, mice—any organism that engages in sexual reproduction. Schwartzman, like the other F_1 gaian researchers, is by no means a clone of Jim Lovelock. Each F_1 has been shaped by a scientific heritage that comes from other lineages, as well. Thus each has a distinct take on what geophysiology is all about, and each contributes a unique combination of interests and talents.

During the 1996 conference on Gaia at Oxford University, I asked five of the F_1's to come together for conversation during a lunch break. All have publications directly relevant to Gaia theory. Two have coauthored scientific papers with Jim Lovelock. Three have worked together in various combinations. I also invited a sixth to join the conversation. He is the first of the F_2 generation. Currently a graduate student, his scientific perspective is being shaped almost entirely by a gaian outlook: He is working closely with Jim Lovelock and one of the F_1's is his research adviser.

The cast, in order of appearance:

TIM LENTON, F_2, age 22, graduate student at the University of East Anglia in Norwich, England.

DAVID SCHWARTZMAN, F_1, age 52, biogeochemist, professor at Howard University in Washington, D.C.

TYLER VOLK, F_1, age 45, earth system scientist, associate professor at New York University.

LEE KLINGER, F_1, age 41, ecologist, staff scientist at the National Center for Atmospheric Research in Boulder, Colorado.

ANDREW WATSON, F_1, age 43, earth system scientist, (then) research scientist at the Marine Biological Association in Plymouth; (now) professor at the University of East Anglia.

LEE KUMP, F_1, age 37, geochemist, associate professor at Pennsylvania State University.

I invited these six because they are all deeply involved in gaian research. The five F_1's are on the board of the just-born Geophysiological Society. Moreover, because they all know one another, I am hoping to extract some rather personal comments out of them. I am curious not only about how each became involved in this research—what attracted them to this new, and still-controversial, science—but what this kind of understanding has done for their own worldviews, their personal philosophies.

Tim Lenton volunteers the first story. "In my secondary school years I was generally bored and disillusioned by most of my teachers, so I read a phenomenal amount of popular science. Jim's book was one of the things I came across. I think it actually might have been my dad [a plant hormone biochemist] who suggested it to me, which is nice. Anyway, I found the Gaia hypothesis to be a dramatic and exciting idea, but it also made sense to my young mind at the time. By the end of my first year in an undergraduate natural science program at Cambridge, I was absolutely determined to do something meaningful. I wrote a letter to Jim saying, 'Look, I'm determined to work on Gaia theory after I get my degree. Can you point me in the direction of any useful people?' Jim wrote back that I must talk to Andy Watson and also come down to Coombe Mill [Lovelock's home and laboratory in Cornwall] and have a chat. I did both. We hit if off right away, and so Jim became my mentor and Andy eventually became my graduate supervisor."

For his doctorate degree, Tim Lenton is developing a geophysiological model of nutrient (that is, nitrogen and phosphorus) regulation coupled with oxygen regulation. Lenton is thus tackling the scientific puzzle identified in 1958 by Alfred Redfield.[43] Why is it that the ratio of nitrogen and phosphorus in the deep ocean is the same as the ratio of the two nutrients required by organisms living in the surface, sunlit zone? And why is it that the amount of oxygen present is exactly what is needed to respire the biomass produced from these nutrients? At our luncheon gathering he is a bit anxious about the day to come, in which he alone will be presenting a talk on the current expression of Gaia theory, written with his mentor, in the presence of a host of accomplished scientists.

David Schwartzman speaks next, his Brooklyn accent a marked contrast to the Queen's English we have just heard. "It all starts when I was about twelve. I grew up in a household of left-wingers. There was a lot of Marxist literature in the house. My uncle downstairs had J. B. S. Haldane's books, and I remember vividly Haldane's account of some of the Russian work on biological weathering. A year later I read *Dialectics of Nature*. So I got a dialectical view of the world early on, which I think Gaia has elements of—interactionism, the whole and the part, and so on. In high school I got into chemistry and then majored in geochemistry in college. I discovered Vernadsky while an undergraduate at City College. Because his book on the biosphere had not then been translated into English, I read his 1945 *American Scientist* paper.[44] That enlightened me on the interaction of biota and the inorganic world. Vernadsky's concept that life is a geological force really impressed me."

Schwartzman continues, "My graduate research at Brown was

in hard-rock geochemistry, doing potassium–argon dating. Later, at Howard University, I got into environmental stuff, particularly looking at lichens as biomonitors, as indicators of pollution. I read Lovelock's book on my second honeymoon. That marriage didn't last, but my interest in Gaia did."

"Because you already knew Vernadsky's work," I interject through the laughter, "was it no big epiphany for you to read *Gaia*?"

"It reawakened what I had read before."

David Schwartzman's Gaia-related papers include one with a title too good to paraphrase: "Biotic Enhancement of Weathering and the Habitability of Earth," which was published in *Nature* in 1989, with Tyler Volk as coauthor.[45] As explained earlier in this chapter, Schwartzman and Volk contend that life played a crucial role, and one that extended for several billion years, in moving Earth's climate away from an early hothouse to the ice-age conditions with us today. Schwartzman is now doing empirical work with lichens and their effect on weathering. At the time of our conversation, he was well into the first draft of a book manuscript, *The Self-Organizing Biosphere*, to be published by Columbia University Press.

The conversation continues as Schwartzman's coauthor, Tyler Volk, offers the next story. "The first time I heard of Lovelock was through *The Whole Earth Review*. It was probably still called *CoEvolution Quarterly* in those days. I believe it was the article about Daisyworld—the lead story; I remember pandas and zebras on the cover.[46] I was reading that journal a lot then because it would have articles about Buckminster Fuller and Gregory Bateson, on whole systems. Anyway, this was while I was in grad school at New York University, doing work on the marine carbon cycle. The Gaia idea fitted into my general interest in whole systems. I went on to read Lovelock's 1979 book and then got more and more into it over the years."

In addition to the biotic weathering papers published before and then with Schwartzman, Volk has published two in which he computes the global cooling triggered by evolutionary innovations. One examines the role of angiosperm plants in Cenozoic cooling; angiosperms may be more effective weatherers of silicate rocks than are their gymnosperm predecessors. The other examines the proliferation of microscopic plankton with carbonate shells, which shifted the locus of carbon burial further offshore. Volk has also coauthored two papers with me. The first presents what we call "the closed biosphere paradox," discussed in the previous section. The second is a review paper of the problems Gaia theory encounters among evolutionary biologists.[47] Volk's book, *Gaia's Body: Toward a Physiology of Earth,* is scheduled for publication about the same time, and with the same publisher, as this book.

The next story is Lee Klinger's. "Actually, Gaia theory developed for me empirically, based on my fieldwork with peat bogs. I had already put together most of my idea of the coupling mechanism between peatlands and ice ages. One winter I went on a ski trip up in Jackson, Wyoming, and ran into an old girlfriend who was manager of a progressive bookstore there. And she said, 'Lee, there's a book you should read by James Lovelock on the Gaia hypothesis.' I said, 'The what?' I had no knowledge of it. So she gave me a copy. Gaia theory awakened in me something I knew was profound. I was thinking, 'My God, the implications of this are incredible.' I remember reading that book with both excitement and disappointment. Excitement that, yes, this is it! But at the same time I felt that I could have come up with some of it—surely not all of it—but some of it on my own. My view of a self-regulating Earth at that time was just a crude organic carbon cycling machine, whereas Lovelock's Earth was an elaborate biogeochemical factory. My Earth was barely alive; Lovelock's was fully alive. I was humbled. At the same time I could see there was plenty of room for my ideas still, because there was no ecology in it yet. It's been a marriage ever since."

Bog Man Klinger (so named by *Discover* magazine in an article about his work[48]) is best known among gaian researchers for his hypothesis that the spread of peat bogs works a positive feedback effect that can turn a cooling trend into an ice age. He spends a lot of time in the field, in bogs ranging from the tundra of arctic Alaska to the jungles of central Africa. Several of his papers challenge conventional ecological wisdom: Forests are not the climax ecosystems, Klinger maintains; bogs are. The only thing that prevents bogs from taking over the planet is the frequency of fire, wind storms, landslides on steep slopes, glacial scouring, and other disturbances that set back the successional clock.[49] Currently, he is investigating whether any of the hundreds of different kinds of volatile chemicals emitted by terrestrial plants into the atmosphere may serve as a kind of gaian-scale hormonal system.

"I am completely different from all of you because I had no idea what I was getting into." Andy Watson, the F_1 with the oldest pedigree as a gaian scientist, begins his story. "In 1974 I was working on a physics degree at Imperial College [University of London], and I wanted to do astronomy. I was just finishing up school, and Jackie and I had started dating [Jackie is his wife]. She wanted to do an undergraduate degree at the University of Reading. So I looked in the prospectus for courses in the Physics Department. I didn't see anything I wanted, so in desperation I looked at what the Engineering Department had to offer. I came upon two lines about a visiting professor, James Lovelock, who does cybernetic aeronomy. I thought, well, it sounds interesting—actually, I hadn't the faintest idea what cybernetic aeronomy is. Anyway, just before my final exams I was in the library and read the *New Scientist* article that Jim had done.[50] That changed the direction of my research completely. From then on, I knew there was something to do at Reading that was indeed very interesting. I wrote to the department and they said, in essence, that although Lovelock is a visiting professor, he never actually visits. 'You'll have to go to him.' So I drove down to see him in the middle of the countryside. We got on very well. I was immediately struck that the entire household was quite mad. I remember walking in and they were all wearing medical masks because one of them had a cold. They gave me a mask, too. I recognized right then that something different in my life was about to happen."

Andy Watson earned his Ph.D. in 1978, with Lovelock as adviser. For his graduate research he spent a lot of time burning newspapers (a stand-in for biomass) under controlled conditions of different oxygen concentrations. This was to ascertain the upper limit of atmospheric oxygen concentration beyond which catastrophic wildfires would make life difficult in terrestrial ecosystems. Andy was happy to collaborate with his former adviser a few years later, when Lovelock hit upon the Daisyworld idea and needed an accomplice who was good at mathematics and comfortable with computers. Watson went on to gain his sea legs and to dip into the empirical world of real life when he took a research post at the Marine Biological Association in Plymouth, England. There he joined the team that conducted the first field test of a hypothesis that scarcity of iron is the reason Pacific equatorial waters are biological deserts.[51]

Lee Kump tells the final story. "Well, I guess my path toward Gaia and Jim was different from everyone's because it wasn't empirical and it wasn't predestined. It was forced upon me! I was in graduate school at the University of South Florida in St. Petersburg, and my adviser, Bob Garrels, had developed a friendship with Jim. Jim had sought out Bob to test some of his ideas that had a bearing on geochemistry. Also, he wanted to see if he could bring more traditional geochemistry people into the fold, so to speak. So he would visit Bob every so often in St. Pete. On one of those visits Bob Garrels came up to me, saying, 'I've got Jim Lovelock coming.' (I didn't know who that was.) 'He's written some things about the Archean, about geochemical cycles, that look pretty interesting, but we're convinced that we see the Earth quite differently. Yet we want to be able to talk about this and to model it. So your job is translator.'

"This was 1984, 85. And at that point Jim had some of the early Daisyworld models. Garrels had his own numerical models, the first version of the BLAG model. Lovelock was convinced that you can't capture Gaia in a box model like BLAG. Garrels couldn't accept that. What was so special about Gaia that it couldn't go into a box model? Everything can go into a box model—right? So one of my jobs was to take Lovelock's Daisyworld model of the Archean and reformulate it as a Garrels-type box model—which I found to be very trivially easy. I could see at the time where they might think there were differences, but really deep down, it felt like there was no fundamental difference in their approaches. The gaian-type Daisyworld feedbacks were just a different functional relationship—higher-order kinetics—from what we were used to working with.

"So I did that," Kump continues. "And I think we made substantial progress in those days. I know Jim felt very good about the fact that geochemists were coming around. And we felt very good about the fact that Jim was coming around, and finding some middle ground."

Lee Kump's expertise covers a wide range in the traditional field of geochemistry as well as the newcomer, geophysiology. His paper on the interaction of atmospheric oxygen, wildfires, and carbon and phosphate burial in the oceans is widely cited by geochemists and geophysiologists. Kump is first author (with Jim Lovelock) of a chapter titled "The Geophysiology of Climate" in an edited volume on numerical modeling approaches to understanding climate. The two teamed up again on modeling work, exploring how the different growth regimes of the marine and terrestrial biota might regulate ice ages. Here they begin with the biological understanding that algal productivity is maximized at cooler temperatures (owing to more nutrients in cooler, and therefore better mixed, waters) and that land plant productivity is maximized at higher temperatures. The coupling of these two systems in a geophysiological model offers insight not only on how the biota might modulate an ice age but also on how both land and sea life might do just the opposite during a warming trend, if the mean global temperature ever exceeds 20°C (it's about 15°C now).[52]

Tim Lenton takes issue with Lee Kump's conclusion that mainstream geochemists and gaian researchers have reconciled their differences, whether they are aware of it or wish to admit it. "Gaia explicitly includes life," he stresses. "If you think in a truly geophysiological way, you're trying to fully incorporate the most important aspects of life: the exponential growth potential, the fact that life inevitably alters its environment, and the fact that life is constrained by its environment. On top of that we can add natural selection as an inevitability. Geochemical box models just don't do that. They can't include life in a sense that is realistic. They miss the essential aspects of what life is."

A few months earlier Tim had traveled to the United States. Tyler and I met him for the first time and talked Gaia all evening over dinner at our apartment in New York. It was then that I began to understand how incompletely gaian Tyler and I were, and how well Tim Lenton served as a knowledgeable, passionate, and innovative teacher and advocate of the purest form of Jim Lovelock's vision. He is, after all, an F_2. Beginning with Gregor Mendel, it has been known that a character or mutation that appears in only one of the parents in a genetics experiment may be completely hidden in the first, F_1, generation of offspring. Rather, the gene must be contributed by both parents in order to be expressed in the body. (Rare genetic diseases, such as cystic fibrosis, work in this way.) None of the F_1 generation of gaian researchers participating in this conversation express the Gaia trait in its pure, Lovelockian form. But Tim Lenton is the scientific offspring of Jim Lovelock and Andy Watson, and he seems to have picked up a Gaia gene from both these parents—not to mention his self-made zeal for this way of thinking. Also, he is the only one of the six to have chosen a graduate science program and advisers expressly so that he could pursue his interest in gaian science.

The conversation now shifts away from the science and the stories of research scientists to the realm beyond science—to questions of meaning.

Andy Watson begins. "The debt we owe to Jim is that he took this study of the Earth and put it into a book that many people who are not scientists actually read. And he was brave enough or stupid enough, depending on which way you look at it, to put it in partially poetic terms and to point out its relevance to philosophy—to the big questions everybody asks as they grow up and as they go through life. Like 'Where does this world come from? How is it that we've managed to be here at all?' Those kinds of questions. So Lovelock's work enables us to connect the everyday business of doing science with a cosmic dimension. And that's the thing that I think is his great contribution. It's certainly the thing that keeps me ever fascinated by this work. I use Jim's outlook to relate what I'm doing in the cosmos in a way that, if I were studying the number of legs on insects, for example, I just couldn't do."

"I also just like the new perspectives that I have on nature," offers Lee Kump, "the questions I would never ask if I hadn't thought along the lines of geophysiology. Tyler, you've probably expressed this better than I can, that is, the 'hypothesis generator' aspect of Gaia. And it's not just the fact that life is involved, either. At the previous Gaia conference [1994, also in Oxford] I talked about subduction zones; there are questions you can ask about subduction zones and the metamorphic reactions that take place in there that you'd never ask as a metamorphic petrologist. But if you think about it from a systems, feedback point of view, there are some interesting things to study that traditional people would never think to study. Gaia makes science more interesting for me—and not just because I can link it into something grander. Because of my early interaction with Garrels and Lovelock, I started out as a top-downist, in the newer generation of people who could start their careers in that way. We didn't have to move from being leg counters to global thinkers."

"I'd like to go back to an issue Andy raised: the cosmic frame," says David Schwartzman. "One of my earliest attempts to tie Gaia into one of my long-term interests, SETI [the Search for Extra-Terrestrial Intelligence], was the possibility that life would perpetuate a biosphere to the point where intelligence could emerge. Of course, that was of vital interest to the SETI people. I went to a SETI meeting in Estonia in 1981 and actually invoked Gaia as a mechanism."[53]

Schwartzman elaborates on the SETI problem and how gaian-like regulation on alien planets might enhance the likelihood of extraterrestrial intelligence. Then he shifts the subject. "I just want to touch on one other thing. That is the aesthetic or emotional aspect of Gaia. Lichens for me became a tremendous aesthetic experience, to tell you the truth. I discovered them as biomonitors and then went out to look at them just to appreciate their global significance in weathering. Here is this inconspicuous organism that geologists like to avoid when they collect rocks. Yet lichens now have a resonance with me."

"I have a question related to that," I break in. "People who are into God can, for example, relate to Jesus. Jesus is something closer, more familiar and human-scale, than The Big Thing. David, you have lichens as your way of communing with the biosphere. And Klinger,

I guess your totem organism is sphagnum moss, or more broadly the bog ecosystem."

"It's a window, sure," says Lee. (The "favorite organism" dialogue with Klinger in chapter 3 took place after this group conversation.)

"Does that sort of relationship with an organism affect anybody else here?" I ask. "Does anybody else have a . . ."

"a favorite organism?" someone offers with a snicker.

"Well, Jim Lovelock likes trees, obviously," says Lee Kump. (Everybody here knows that Jim and his family planted thousands of trees on their property in Cornwall and that he laments the loss of so much of Britain's original forest.) Kump continues, "I think Lynn Margulis's and a lot of people's favorite is the microbial mat community. It's probably the best microcosm for studying gaian feedbacks."

"One more thing," I venture. "Is Gaia the sort of thing that everyone here finds empowering? Does it make you feel at home in the universe? Did anybody feel as if something were missing, philosophically, until you discovered Gaia, and now you feel whole?"

Tyler Volk begins. "Learning about the science of Gaia makes me feel a lot more connected with the Earth. When I go for a walk in the woods now or just look up at the clouds, I feel like I'm inside a giant metabolism."

Tim Lenton responds with a story. "I was quite a postmodern kid and had a real meaning crisis around the age of five. I got trapped into watching television and the whole consumer thing; I had a lot of attachment to small objects, possessions. All the time there was a painful sense of a lack of meaning in the current social beliefs, the dominant way of seeing and thinking about things. And then I had a variety of experiences that would later fit into what I came to learn was the gaian worldview."

"What sort of experiences?" I ask.

"One that comes to mind immediately happened when I was about eight years old," Tim responds. "I was on holiday with my family in Switzerland. We were bathing in a mountain lake in bright sunshine. Suddenly my mum called out that a storm would be upon us within ten minutes. The rest of us laughed, but she was right. I vividly remember drying off and dressing as the storm moved in, then being soaked to the skin and stung by hailstones as we battled our way to shelter. As I recall the feeling now, it was an experience of great humility. The human barriers of clothing were broken down, and I was right back in touch with natural processes. Though painful, it was also liberating, and it gave me a sense of being a tiny part of a larger system."

"In a sense, Connie, you're asking, 'Do we have a religious feeling about Gaia?' " Andy Watson now has our attention. "I have gone through that phase. I don't think that I do anymore have a particular religious feeling for Gaia. But what a gaian understanding has helped me with is in seeing ourselves as latecomers, as observers on this planet and in a greater cosmic scheme. To see that there is in some sense—albeit very halting, albeit as much backward as forward—that there is a direction from one scale of complexity upward, and that currently, like it or not, we are the most complex things we know about, and to want to know what's beyond that. Gaia has helped me to see the sort of project for humanity and for the universe that we are part of. If that's religion, then, yes, Gaia has had a powerful influence on me."

"For me," says David Schwartzman, "it's realizing that your atoms will re-enter these cycles. It's a form of immortality for a materialist."

"I don't want to come back as a nematode," protests Klinger.

"Don't worry," Schwartzman consoles, "you'll be converted to something else eventually."

"Ah," warns Watson, "first you'll have to spend twenty, thirty thousand years in the abyss—that's where all the carbon is—before you'll get to come back as a nematode."

Our little group adjourns and cycles back into the larger reservoir of scientists, who are now assembling for a group photograph before the afternoon talks begin. I was grateful that these research scientists took the time to meet with me and tell their personal stories. And I was encouraged that they were willing to say anything at all beyond the science. Questions of meaning are not normally encountered in scientific gatherings, or in the day-to-day work of research science. But there was one scientist in attendance at this conference for whom communicating the meaning of gaian science is the core of his job description.

I decided to track down Stephan Harding.

Beyond the Science

Stephan Harding was invited by Jim Lovelock to present the results of his Daisyworld modeling efforts at the 1996 Gaia conference in Oxford. Like some of the other scientists at this conference, Harding has published scientific work with Lovelock.[54] Harding's modeling of a food web of organisms that both affect and are affected by their environment is yielding tantalizing clues that more diverse communities may, in fact, help stabilize environmental conditions. The results also indicate that herbivore-induced patterns of plant biodiversity have corresponding effects on the climate of an ecologically complex Daisyworld. Nevertheless, for Harding, "the whole reason for gathering scientific information is to provide a cognitive basis for developing wide identification with nature."

Harding, age 43, credits this nuancing of science to another mentor: Arne Naess, the Norwegian ecophilosopher who originated the worldview of deep ecology. "The aim for me of having all this information," explains Harding, "is so that I can feel a greater sense of belonging and rootedness in the life of this Earth. This is identification. It gives me a terrifically satisfying sense of joy in being part of such a wondrous planet with its beautiful ecology. And out of that feeling comes the conviction that life, both human and nonhuman, has intrinsic value—value in itself. So I have this strong inner sense that the nonhuman world matters and must be protected, irrespective of its usefulness to humans—and even to Gaia itself.

"The more you know about someone you like," he continues, "the more you identify with them—right? For example, I know that my grandfather escaped from a Russian camp and then spent three years walking to freedom in Poland, so I think Wow! that was my grandfather. And then if I know about the experience of my great, great, great, great grandfather—which is, so to speak, the history of the planet itself—then I feel at home on the Earth. That for me is the sole aim of accumulating all this scientific information."

Stephan Harding is staff ecologist for Schumacher College in Dartington, Devon. This small, alternative college in the English countryside—named in memory of E. F. Schumacher, author of *Small Is Beautiful*—was founded in 1991 upon the twin convictions that the dominant worldview of Western civilization has serious limitations and that a new vision is needed for human society, its values, and its relationship to the Earth. In Harding's, words, "The mission is to look deeply into the causes of the ecological crisis. What went wrong, and how can we make amends? What sort of inspiration is there in our own culture, and in other cultures, that can help us find a better path?" One- to three-week courses are offered at Schumacher on a range of topics, with credits transferable to the home institutions of those participants working on masters degrees. After the Oxford conference, Harding must prepare for a Gaia course he will soon be team-teaching with guest lecturer Lynn Margulis.

Harding took his Ph.D. in field ecology at Oxford University. His research topic was the ecology of muntjac deer, a genus of small deer native to Southeast Asia. (Recall the story of his attraction to this creature, presented in chapter 3.) Fresh out of graduate school, Harding went to Costa Rica, where he taught ecology. Returning to England, he heard about the then-forming Schumacher College. "I'd always had a deep identification with other life forms and ecosystems. That was one of the things that frustrated me about science, as that kind of identification was never developed or brought into science. So I was unhappy overall with science, really. I loved it, but I was looking for something else, something more multi-dimensional. Schumacher College brought me closer to where I wanted to go to. And there I met Jim Lovelock, as he gave the first course upon the opening of the college. We got on really well. He said, 'Oh, you're an ecologist. You might be interested in what I'm doing.' I certainly was! I got involved and started working on Daisyworld models, making them as ecologically realistic as possible."

Even during sessions in which Gaia is not the focus of the Schumacher offerings, Harding always gives at least some talks on Gaia and also on ecophilosophy or deep ecology. "Most people who come to Schumacher have already heard something about Gaia. They're familiar with the name but don't know quite what it is. And they don't realize that it is a scientific theory. So when they hear a scientific presentation of it, they are amazed. They really love it because they begin to have a rational basis for understanding what's going on. They begin to realize that the planet has life-like qualities of self-regulation. They begin to grasp the vast time scales involved. And they begin to see where humans fit in it all."

He continues, "Beyond that, some people come to see that the gaian view releases us from having to be the planet's stewards or controllers. This brings a tremendous sense of relief, as if a great burden has been lifted. At last, we are free to be what we are—one of a multitude of species, an equal partner in Gaia's ongoing evolution.

"And that's the understanding people want," he concludes. "That's what people want. That's what *I* wanted. The only reason I went into science was I was looking for that: a way of using science to foster a sense of wide identification with nature. After all, when you're a kid—when I was a kid—I completely identified with wild things, nature. Adoration, love, happiness. Then I discovered there's a science called ecology. The prospect of learning ecology excited me because I thought that more knowledge would help me identify with nature even more. Unfortunately, the way we're taught science, the opposite happens. The more we learn, the more separated we become—the more objective."

Harding believes the alienation born of an overemphasis on objectivity and measurement is deep-seated in science, going back to Galileo, but it needn't be that way. "Our spontaneous experience of nature—colors, sounds, smells, textures—these are primary," he contends. "Scientific theories and the world of abstraction obviously come out of all this rich, sensual world. It is my hope that gaian science will become part of a rediscovery of a science of *qualities* that was developed by Goethe and that Brian Goodwin and others are pursuing today."

Shifting topics from the conduct of science to the teaching and interpretation of science at Schumacher College, I ask Stephan Harding to comment on a question that has become more than idle curiosity for me of late. I have a sense that people in general are more easily drawn into a deep affinity for Gaia than for the evolutionary epic. I know Thomas Berry has taught Schumacher courses on the Universe Story, and so I ask Harding about the differences in participant responses to Gaia and to evolution.

"In my experience it's much easier to move people if you begin with Gaia," he responds. "I think it is best to explain first that the planet as a whole is integrating, self-regulating, self-correcting—even alive—and that it has been so for thousands of millions of years. You start off from there. And only then do you teach the evolutionary story. The story makes much more sense that way, because now it can be seen in the context of what it is that has been evolving: this global-scale entity called Gaia. It's the psychological contact with the emergent property of self-regulation that does it: the sense of something new emerging from the parts. That's what it is. That's what people connect with."

As to Harding's own spiritual grounding, "My main source is the gaian perspective. Before I had that, I had no way to tie together my spontaneous love and awe for nature with my rational knowledge of science. Gaia, for me, makes sense of the whole thing. It gives everything a context, a continuity. It provides an overarching connectedness. Ecology and evolutionary biology are very important to understand and, from a spiritual sense, are very useful. But how do they tie in? What is the bigger pattern of which they are part? Gaia fulfills that. So that's why, for me, Gaia theory is primary."

Gaia theory is primary for me, too. But so is the evolutionary epic. I couldn't place one as more fundamental than the other. Gaia gives me a sense of green space. The evolutionary epic gives me green time. In turn, Gaia has an evolutionary history. And the evolutionary epic, the grand pageant of organismic forms and talents, forever and always takes place within the embrace of a nurturing biosphere.

I will, however, accept a celebration of biodiversity and bioregionalism as somewhat derivative byways of the spiritual path of science. Biodiversity is, after all, the current expression of the ongoing creation story I live by: the evolutionary epic. My bioregion is the small patch of Gaia I have come to know and love and that I hope to protect.

The flux of my life has required shifting allegiances among bioregions and the life forms therein—from the Great Lakes and monarch butterflies of my childhood, to interior Alaska and tundra mosses, to Southeast Alaska with its enormous slugs, to Cascadia and its coastal kelp forests, and now split between the riot of deciduous trees in the lower Hudson River Valley and my beloved cottonwoods of the Upper Gila Watershed. Throughout, I have never once set foot outside the domain of Gaia: I am a homebody. And I have never once bonded with a creature whose story is not that of the evolutionary epic.

As an avid amateur naturalist, I know that a feeling for biodiversity and a feeling for an ecology of place do not require scientific understanding. Awareness of the fruits of science enhances those feelings enormously, but a grounding can nevertheless be achieved experientially. For the evolutionary epic and Gaia, however—ah, that is another matter entirely. At these largest scales of time and earthly space, the sophisticated tools and the pooled minds of the scientific community are essential. Perhaps that is why coming upon the science of the whole Earth and the science of the history of life has been so memorable, even epiphanic, for those of us lucky enough to have encountered these subjects in deep and meaningful ways. That often means through the written words of visionaries. Visionaries are those who become possessed by the grandeur of what they have glimpsed and who possess gifts to convey that grandeur to others—to *evoke* that grandeur in others, as Mary Evelyn Tucker reminds us—and to do so with grace and with an openness that allows the rest of us to peer into their very souls.

Recall Lee Klinger's excitement upon reading Lovelock's *Gaia*, after a ski trip in Wyoming. Recall my own re-enchantment with biology that ensued from a chance encounter with the same book. Recall Brian Swimme's devotion to the universe that blossomed upon reading

Teilhard de Chardin in a Jesuit high school and, years later, upon encountering Thomas Berry's interpretation. Recall my own enthusiasm for the epic that Julian Huxley called forth.

I can now turn to the evolutionary epic for a creation story that is at least a match in its grandeur with that of any other people anywhere in the world. I can look upon other species truly as kin and come to know through science some of their stories—which of them, for example, are elders and which are newcomers, like me. I can look for intimacy of place in my bioregion and yet still know that wherever I may roam, I will always be home. Like Dorothy in Oz, I discover that I can be home in an instant. But I keep my eyes open wide, face lifted toward the sky, and recite a very different mantra: "like a cat's fur, a bird's feathers. . . ."

The biosphere is awesome, self-regulation or no. Shall we call it Gaia, whatever the outcome of the scientific debate? Why not? I like to think of Gaia not as an entity like an organism, but more as a swarm—the kind of distributed intelligence evident in an ant hill and that complexity theorists watch as pixels blink on and off. I have a dream that one day the biosphere will be well enough known and venerable enough a topic that the question of *what* it is will no longer arise. Lee Klinger had already come to that view several years ago when he scolded me, "A cell is a cell; an organism is an organism; and Gaia is Gaia." Gaia theory will have come of age when we no longer strain for a metaphor to explain it, but when Gaia itself becomes a metaphor. I envision an nth edition of *Webster's* with this entry:

> **gaian** *adj* **1:** self-sustaining **2:** smoothly functioning, dependable **3:** powerful, worthy of respect **4:** *slang* : very impressive.

For Gaia and the evolutionary epic, therefore, I have found that the way of science can be as alluring as it is essential. Bioregionalism and biodiversity, however, crucially contribute the human-scale sensory contact with the pulsing world; each offers us primates a vine to scamper across to the two great realms of green space and green time that can be reached, ultimately, only by way of science.

Meaning-Making

Meaning is always relational. Something is meaningful to that which beholds the meaning. Meaning is therefore a subjective experience, but this does not in any way make it illusory or restrict it to the province of human observers or participants. Recall Stephan Harding's statement in chapter 3 that to a muntjac deer, a single bush can be meaningful, "a juicy, succulent, significant item." Few would deny that the muntjac deer has the capacity to interpret its environment in meaningful ways—discerning things and situations that it regards as highly favorable from those that are outright dangerous or merely benign. Ursula Goodenough and many other biologists contend that meaning emerges in the cosmos with the very first living cell, which, by definition, *strives* to maintain its own existence.[1] It has a telos, a project. Its reason to be is to be and to produce more of its own kind.

That the world is meaningful to even the smallest and simplest forms of life is particularly evident among organisms that can move. Motile cells will move toward favorable thermal and chemical conditions and away from the reverse. Even cells lacking such talents are able to interpret favorable and unfavorable conditions, stoking up metabolic and reproductive equipment in the former and hunkering down in a quiescent state during the latter. Heat and acidity *mean* something to microbes, and they respond with vigor.

"A rock conserves no identity," writes philosopher Holmes Rolston. Organisms, in contrast, "post a careful, semipermeable boundary between themselves and the rest of nature; they assimilate environmental materials to their own needs. They gain and maintain internal order against the disordering tendencies of external nature. They keep rewinding, recomposing themselves, while inanimate objects run down, erode, decompose."[2] Thus an exclusively science-based view of the world does not attribute self-interest and responsiveness to a rock, a mountain, a river—although all of these inanimate presences may be enormously meaningful to organisms who depend on them. Meaning, rather, emerges with life.

From a gaian perspective, however, we know that every atom of hydrogen in every molecule of water now sloshing through the world's rivers and oceans or sequestered in polar ice has almost certainly been drawn into living tissue a score of times since the first plant composed itself out of water, carbon dioxide, and sun. Similarly, some of the constituents of all but the most ancient rocks may have at some time been bound within a body. The atmosphere as a whole is not alive; it has no self-interest. But "like a cat's fur, a bird's feathers, or the paper of a wasp's nest,"[3] it is the product of a living system. Scientific instruments can sense the annual rise and fall of carbon dioxide in the land-rich Northern Hemisphere, a cycle which scientists themselves have called "the breathing of the biosphere."[4] Gaia theory thus reanimates the inanimate.

Meaning emerges with life. And meaning enriches with evolution. It even reaches beyond the confines of Earth. Photons sent to Earth by the sun became enormously meaningful to the first photosynthetic form of life. Diurnal cycles invented by various life forms are tributes to the meaning of night and day. The sun itself became meaningful to the very first plant able to grow toward the light and the first animal that used solar orientation (as honeybees do today) to remember how to find a patch of something meaningful that was worth revisiting. The tidal pull of the moon has long been vitally meaningful to the many marine hydroids and mollusks and echinoderms and arthropods and vertebrates that must synchronize their spawning. Polaris, the North Star, is likely very meaningful for birds who use the stars to calibrate their internal magnetometers essential for long, nocturnal journeys. The whole expanding cosmos is meaningful to Brian Swimme, as are its ghosts of ages past—from the former star Tiamat to the pterosaurs of the Cretaceous.

To find the universe meaningless is as much an interpretive act as to find the universe meaningful. When a scientist, for example, Stephen Jay Gould or Steven Weinberg, declares the universe or the evolutionary story to be pointless, we thereby learn more about the psyche of the claimant than about the cosmos. It is a category error to expect such valuation—one way or the other—to come cleanly from the realm of science. Rather, such judgments are an interpretation drawn out of science, culture, and lived experience. Bryan Appleyard wrote a whole book based on the premise that Gould and Weinberg could indeed make proclamations of cosmic meaninglessness with the full weight of scientific authority behind them. Appleyard therefore judged science to be deeply troubling for "the soul of modern man." He presents Gould not as one who, like Thomas Berry, extends science into the realm of meaning, but as "an elegant defender of the hard truth of science."[5]

Appleyard is wrong. The hard truth of science is that this universe is billions of years old, that complex atoms were forged in the core of stars, that Earth is younger than the Milky Way Galaxy, that life evolved from the simple to the simple and the complex, that dinosaurs and ground sloths once roamed this planet, that humans are newcomers to the pageant of life. The hard truth of science says nothing about the meaning of all these facts.

Meaning is also too often confused with *purpose*. One can find meaning in the pageant of life—indeed, in life generally—without necessarily declaring it all has a purpose of some sort. The evolutionary epic need not be on track toward some preset goal, some greater good, in order to be meaningful to you and me. Moreover, to assume that for the universe to be meaningful today it must have possessed a meaning unto itself or to an other (such as a creator-god) right from the outset is a very constricted notion of meaning. By that standard, your own life today is no more meaningful to you or anybody else than it was the day you were born. Such an attitude exemplifies the lingering power of the Biblical creation story, which still imbues our thinking. To make the leap to a fully evolutionary outlook, we would come to realize that *everything* in the cosmos emerges through time. Light emerges, atoms emerge, molecules emerge, galaxies emerge, life emerges, vision emerges, flight emerges, frustration emerges, terror emerges, joy emerges, compassion emerges. So, too, has meaning emerged.

> I can hear the sizzle of newborn stars,
> and know that anything of meaning, of fierce
> magic is emerging here. I am witness
> to flexible eternity, the evolving past,
> and I know we will live forever,
> as dust or breath in the face of stars,
> in the shifting pattern of winds.
>
> —*Joy Harjo*[6]

Despite the inherent subjectivity, meaning-making by humans is not mere fabrication; meaning is not an arbitrary, infinitely malleable construction out of nothing. It is a response to, a declaration of relationship with, Earth and the cosmos. To find meaning in the cosmos is no less legitimate than to have an aesthetic response to a landscape. Others may have a different response, but to be fully human is to have a response of some sort.[7]

All human cultures have cosmic stories that impart meaning. These stories transmit cultural wisdom of how things are, how they came to be, and what things matter.[8] They provide the ground for constructing value and for envisioning our own human role in the cosmic enterprise. Tell me a creation story more wondrous than the miracle of a living cell forged from the residue of an exploding star! Tell me a story of transformation more magical than that of a fish hauling out onto land and becoming amphibian, or a reptile taking to the sky and becoming bird, or a bear slipping back into the sea and becoming whale! If this science-based culture, of all cultures, cannot find meaning and cause for celebration in its very own cosmic creation story, then we are sorely impoverished indeed.

What meaning can we then make of the scientific story? More broadly, how can an understanding of the evolutionary epic, of biodiversity, of bioregions, and Gaia affect our psychological states, our commitment to credos of ultimate value, our sense of our own role on Earth and in the cosmos?

A Federation of Meaning

The great metaphysical binaries that have long served as contemplative aids can be called up from the sciences, too.[9] Being and becoming, passivity and agency, communion and individuality, the Many and the One—these and more can be drawn from the range of scientific disciplines that foster understanding of the evolutionary epic, biodiversity, bioregions, and Gaia. Yin and yang, these binaries are not so much opposites as complements.

Sometimes, for example, it may be more helpful to contemplate the arrow of time than the cycle of time, or the diversity of life rather than its unity. On other occasions, the reverse holds. Winter, with its biological slumber, is the season for putting one's faith in the eternal return: the cycle. In the darkest months we take comfort in knowing that sun and greenery will indeed reappear. In contrast, spring is the season of adventure, for cutting loose and setting off toward the unknown. In spring we savor freshness and experiment, but by doing so we open the door to the risk of setback as well as the prospect of improvement.

The evolutionary epic is first and foremost a celebration of the arrow of time, an irreversible parade of biological novelty upon novelty. More, it is a tale of enrichment. Unicellular life gives rise to multicellular life; life in the sea creeps out onto the land; prostrate land plants explore the vertical and forests are born; insects and pterosaurs take to the air, followed by birds and much later bats. Throughout, animals take on every color of the rainbow to attract a mate or to broadcast toxicity; later, flowers do the same to woo pollinators. Teeth and nematocysts, alarm calls and songs, bark and bur: All these innovations yield an escalating torrent of creativity.

Gaia, in contrast, evokes the power of the cycle, the ceaseless spinning of nutrient cycles that keep this planet fit for life. A plant extracts carbon dioxide from the atmosphere and binds it in carbohydrate; a fungus or microbe or animal later feeds on that molecule and returns carbon dioxide to the sky. Photosynthesis, respiration; photosynthesis, respiration—eternal movement and eternal return. Similarly, nitrogen is carried by life through an endless web of conversions from the gaseous dinitrogen molecule that accounts for most of the atmosphere to nitrate, ammonium, protein, and back again.

Gaia as the cycle of time, evolution as the arrow of time: these are the obvious pairings. But like the oppositional dots of dark and light in a yin–yang symbol, Gaia may also evoke the arrow, and evolution the cycle. For example, by taking a Vernadskian perspective of a biosphere that has generally grown bigger, faster, and smarter through the eons, Gaia is seen to pair with the arrow of time. If David Schwartzman is right about a long-cooling Earth, then the arrow of development is again the image called forth. Gaia is, assuredly, homeostatic over the time scale of humans and redwood trees. But over the grand sweep of evolutionary time, trends may appear. Gaia is thus the allure of becoming as well as the sanctuary of being.

So, too, the arrow of evolution curves back on itself to become a cycle. There is repetition in the epic as much as invention. Today, well-watered lands with moderate climates support rich forests of angiosperm trees. In the Mesozoic era, the gymnosperms reigned. Before them, the niche of towering photosynthesizer was filled by lycopsid trees (whose modern descendants are diminutive club mosses on forest floors) and by sphenopsid trees—ancestors of *Equisetum*, or horsetails. Turning to the ocean realm, we learn that reefs were built first by calcareous algae, then by sponges, then by rudist clams. At the end

Cretaceous, the rudists gave way and corals claimed the reef-building niche, which they have carried on through today. One might look upon that sequence as algae to sponge to mollusk to coelenterate. But one might also interpret it as reef-builder, reef-builder, reef-builder, reef-builder. There is a cycle of attainment and replacement, attainment and replacement. Similarly, at the end Cretaceous when Earth was "struck by a shooting star,"[10] the dinosaurs were not to return. But various mammalian lineages, including that of the modern rhinoceros, evolved into the form and function of tanklike herbivore, echoing *Triceratops*. Tall brontotheres and, later, giraffes stretched their necks into the trees as the brontosaurs had done tens of millions of years earlier. One group may have laid eggs and the other suckled young, but in form and ecological function the players past and present are remarkably the same. More broadly, the arrow of evolution moves not just from sea to land but cycles back to the sea, in the form of eel grass and whale, and earlier, mosasaur and ichthyosaur. From land to air to land is the shared path of the lineages that led to ostrich, emu, dodo.

Within the evolutionary epic other binaries emerge, beyond the arrow and cycle. Cleavage, or differentiation, among lineages—owing to mutations and genetic reassembly, judged by natural selection—creates the very branches and twigs of the tree of life. Early on, however, some microbial branches merged, yielding the integrations of the amoeba cell and the algal cell. Still later came the integration of unicellular life into multicellular life, and thence a differentiation of cells into distinctive organs of animals and plants, followed once again by an integration of individual animals into social groups. Differentiation and integration: These are complementary principles driving evolution.

Similarly, strife and synergy spill out of the evolutionary epic. A forest is a dramatic place to contemplate this binary. Looking up, one can witness the results of a savage "arms race" of competition for sunlight that long ago made trees into towers, driving the redwoods and pines and beeches into a frenzy of skyward longing. How much less consumptive of lignin and cellulose had they all called a truce at, say, a four-foot ceiling![11] Instead, they rise as high as the pull of water transpiration and the thrashing of storms will allow. Below ground, in contrast, lies a kingdom of goodwill, visible to the human eye only when mushrooms push up through the duff to deliver spores to the wind. Below ground pulses a symbiotic partnership of fungal threads delicately probing a network of roots. Soil minerals essential to life are delivered by fungi to the vertical giants. In turn, sun-ripened sugars of the green canopy stream downward, to be shared with the underworld.

Strife co-created tooth of predator with hoof of prey: wolf and caribou, lion and zebra, cheetah and gazelle. It was strife that suggested armor to the armadillo, quills to the porcupine, shell to snail, carapace to terrapin. It was strife that gave keen eyesight to coyote and eagle, night vision to owl. It was strife, in turn, that gave the rabbit its ears, octopus its ink, chameleon its camouflage. Equally, it was strife born of a challenging climate that cloaked rodent and bear in fur and coaxed them to consider hibernation. It was strife that compelled plants to shed their leaves in times of frost or drought. And it was strife in its ultimacy—death—that made room on a finite planet for the sheer excess essential for experimentation, novelty, and hence enrichment.

Meanwhile, the synergy born of mutual aid paired petal with pollinator. Synergy brought algae and fungi together into the hardiest beings of all: lichens. Synergy made fit the evolved partnership of ant and the aphid it milks for honey, the cooperative agreement struck by plant-eating mammals and the plant-digesting microbes in their guts. Synergy shaped the ecosystem, the biosphere. Finally, herds, flocks, swarms, schools, gaggles—all associations of similar beings—trumpet the benefits that come with sociality.[12]

In one sense, evolution pushed by strife and synergy is something that happens *to* a lineage; in another, evolution is something that a lineage works upon itself. There is agency as well as passiveness in the epic. The lineage that led to antelopes evolved fleetness rather than the mass and horns of a rhinoceros because at some crucial juncture, an ancestor chose to flee rather than fight. It was only because an ant took the initiative to prod the abdomen of an aphid that the ant–aphid symbiosis could evolve. And what is the peacock's tail if not the mark of an extravagant willfulness—the sexual preferences of a long line of pea hens? Stephen Jay Gould has celebrated the role initiative plays in evolution: "Organisms are not billiard balls, struck in deterministic fashion by the cue of natural selection, and rolling to optimal positions on life's table. They influence their own history in interesting, complex, and comprehensible ways."[13]

Evolution by way of strife and synergy, initiative and happenstance is a process. Nevertheless, the evolutionary epic—the story of evolution—records the comings and goings of real-life entities, summed to biodiversity. In turn, Gaia is a self-renewing planetary system—an entity—that emerges from the underlying processes, the biogeochemical cycles. Both Gaia and the evolutionary epic are expressions of deterministic as well as contingent forces. Fixation of atmospheric carbon and atmospheric nitrogen surely had to occur for the gaian cycles to emerge fully, but who could have predicted which lineages would take the lead in these innovations? At some point animal life, surely, would find its way onto land and take to the air, but who would have predicted that the arthropods and the vertebrates would excel in these endeavors?

Consider another binary: Do we view ourselves and all the diversity of life as equal members of what Thomas Berry calls the Earth Community? Or do we see ourselves embedded in the greater whole of Gaia? Is our primary devotion thus to our sibs (fellow species) or to a kind of parental entity that we hold in awe? More, can we truly feel at home at the scale of the planet or do we require most the intimacy of bioregional devotion? These all are complementary perspectives that serve our psyches and our societies in different ways and at different times. Joined, both halves of each binary and all the binaries summed yield what Loyal Rue calls a federation of meaning.[14]

In Asia this federation of meaning is called the Tao. The Tao emerges from polarities—not opposites. Like north and south on a magnet, one pole simply cannot exist without the other. Think of the traditional light-and-dark tadpole symbol of Asia, or think of a Möbius strip—a ribbon of paper, twisted once and joined end to end, thus making a circular study of sameness and difference. Cliff Matthews, a chemist and origins-of-life researcher, has created what he calls "a mandala for science."[15] Its traditional mythic elements include the yin–yang symbol and the uroboros. The uroboros began as the circular image of a serpent eating its tail; it is rendered by Matthews as an arrow delineating a cycle. At the heart of the image is a new mythic symbol proffered by science: the spiral staircase of the DNA molecule.

Overall, by way of science we arrive at a deep understanding of the evolutionary epic, biodiversity, bioregions, and Gaia. This understanding, in turn, can be extended far beyond science into the realm of meaning, elucidating the great metaphysical binaries: the cycle and the arrow of time, being and becoming, unity and diversity, differentiation and integration, strife and synergy, agency and passiveness, initiative and happenstance, contingency and determinism, chance and necessity, community and individuality, the global and the local, the past and the future. Both parts of each pair are equally important and equally comforting at different points in our lives.

By way of science we can also green our sense of space and our sense of time. At the most immense scale of earthly space, we awaken

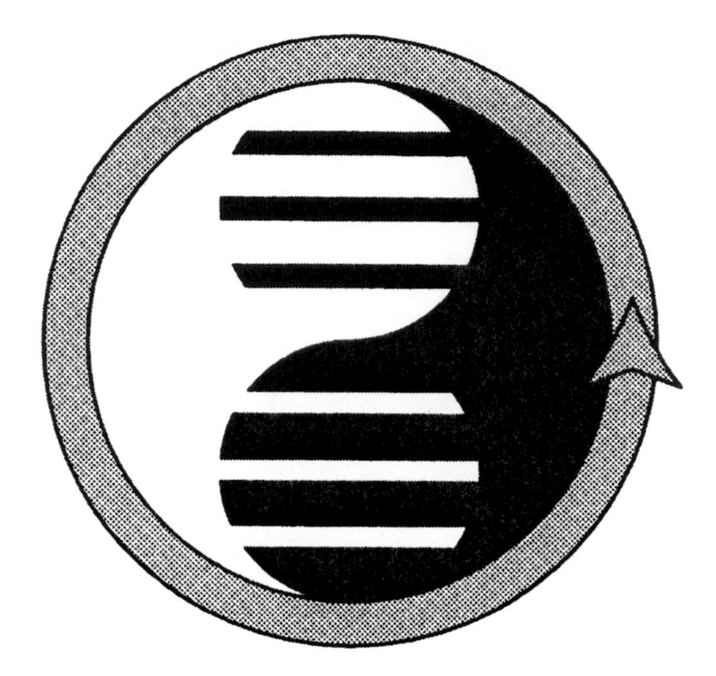

A MANDALA FOR SCIENCE, created by Clifford Matthews. Adapted from Clifford Matthews, 1995, *Cosmic Beginnings and Human Ends: Where Science and Religion Meet* (Chicago: Open Court), p. 26.

to a planetary presence; at the most immense scale of time, we open to the only story that encompasses all peoples and all living beings. More to the human scale of perception, we greet our bioregion and recognize all our planet-mates as fellow travelers in a story that bursts forth in the now with a splendor infinitely enhanced by awareness of what came before and by dreams of what may lie ahead.

Green space, green time—green spacetime. There is no separation. Gaia has persisted for nearly four billion years only because evolution has been building the nutrient cycles and refilling vacancies in ecological niche upon niche. Similarly, at no point in the immense journey of earthly life has the evolutionary epic taken place outside the hospitality and challenges of this self-renewing planetary system.

VALUE-MAKING

Twenty years ago I worked for the president of the Alaska State Senate. Senator Rader's closest friend was a leader of the opposing party. Occasionally I had a chance to listen in while the two (both lawyers) threaded down through one another's positions, challenging logic and efficacy along the way, until at last the philosophical foundations were reached. Usually the positions of both proved fit enough to survive to that point. What struck me was not so much the intellectual beauty of the ratcheting of attack and defense. Rather, I was taken by what happened once the ultimate beliefs were reached. At that point each would acknowledge the validity of the other's position—and that was that. The disputation would end abruptly and the two would happily move on to other topics without the slightest sign of ill will or edginess to get in a final word.

Here is the lesson that has been with me ever since: It is pointless to argue ultimate beliefs; but it is crucial to know what those beliefs are—in one's colleagues, in opponents, and, most important, in oneself.

These bedrock beliefs, in a philosophical context, are known as *ultimates* or *fundamentals* or *axioms*. Philosopher Arne Naess calls them *norms*; William James spoke of *over-beliefs*. Whatever the name, in one's felt experience, bedrock beliefs are not mere propositions. They are self-evident truths that inhabit the heart as well as the mind. An ultimate belief decorates the abode of the heart with items of care, concern, and commitment; the mind, with rational deductions and codes of moral behavior. Here are some familiar examples of ultimate beliefs, articulated as creeds or credos, from world religions and secular societies:

> The Lord our God is one God.
> I believe in God the Father Almighty, maker of heaven and earth.

There is no god but Allah, and Muhammad is his prophet.

That thou art.

I take refuge in the Buddha; I take refuge in the Dharma; I take refuge in the Sangha.

As above, so below.

Nullius in verba. (No faith in words.)

We hold these truths to be self-evident: that all men are created equal.

Liberty, fraternity, equality.

Today's Christian green movement rallies around the credo "The Earth is the Lord's, and we are its stewards." Other green credos of our era include "Earth first!" and a phrase drawn from a poem by Robinson Jeffers, "Not man apart."

Those of us whose bedrock worldviews come in large part by way of science probably don't have an explicit credo that we can recite on demand. We have beliefs, surely—deep and perhaps extremely well thought out beliefs. But how do we communicate those beliefs to others? For this purpose, a credo of one's own design can be immensely valuable—not to mention therapeutic in times of stress or confusion. "This I believe. . . ." Provided that the statement is deep enough,[16] even those who differ will sense the ultimacy of it and back off from the most venomous forms of disputation.

A confession: Beginning this book, I had no articulated credo. I even scoffed at a friend who warned that, for a project such as this, I had better develop one. Along the way, however, I realized I had a long-standing and unswerving commitment to each of the ultimates introduced in the four core science chapters: the pageant of life, the diversity of life, bioregions, and Gaia. From that point, credos quite naturally emerged.

- The evolutionary epic is my creation story, and the pageant of life is evolution's great achievement.

- The evolutionary epic is my creation story, and the diversity of life is evolution's great achievement.
- The evolutionary epic is my creation story, and the richness and integrity of bioregions is evolution's great achievement.
- The evolutionary epic is my creation story, and this self-renewing living planet is evolution's great achievement.

Note first that this is a cosmologically based value system. All the axiomatic statements of value are derivative of the creation story I embrace. That creation story is not identical with the scientific history of life; it is, rather, the evolutionary *epic* as I have personally rendered it, based on the stories told by Julian Huxley, Edward O. Wilson, Brian Swimme, Thomas Berry, Lynn Margulis and Dorion Sagan, Jim Lovelock, Mark and Dianna McMenamin, Vladimir Vernadsky, Andrei Lapo, Alister Hardy, John Maynard Smith, Ursula Goodenough, Loyal Rue, Theodosius Dobzhansky, Jeffrey Wicken, Richard Dawkins, Loren Eiseley, Annie Dillard, and others. Although true to the science, the epic that dances in my soul is a retelling of the strictly scientific story in a way that puts it on a par with mythic narratives that have long motivated human cultures. It is poetic, awesome, inspiring, accessible to my level of understanding, and deeply meaningful.

Theologian John Haught has written, "Religion holds that if values are to have an unconditional claim upon us, they require a transcendent source radically independent of fragile human existence."[17] For Haught and many others, that transcendent source is God; for me, the source is this self-enriching cosmos whose story is the evolutionary epic, as revealed and forever reshaped by science. The evolutionary epic resonates because it tells me what I most want to hear; it aligns with a nascent value system gained through my experience of living. It is nuanced in such a way that the four ultimate values—the pageant of life, the diversity of life, bioregions, and Gaia—are inseparable from the plot and characters. Out of the epic, values upwell as naturally as water from a spring.

This is all a personal response, though much of it is shared by the people I associate with. It is important to remember that not everyone who accepts—even exults in—the scientific portrayal of the history of the universe and life grounds a portion of their value system on this story. Stephen Jay Gould is the most vocal example of a distinguished evolutionary biologist in America who espouses this sort of separatism;[18] John Maynard Smith (whom you will meet in the final section) is Britain's leading separatist. Both are nontheists. Both are keenly aware of abhorrent human-to-human value systems that, not long ago, were grounded on a crude "survival of the fittest" maxim drawn out of Herbert Spencer's works by way of Darwin's *Origin of Species*. This maxim, moreover, wasn't even true to the full understanding of evolutionary science available to that era, as cooperation within and among species was (and still is) known to be just as important a tool of evolutionary change as is struggle.[19]

Crucial here is that none of the four credos I have posited delves into the human-to-human realm at all. For this, I continue to rely on an amalgam of within-species ethics acquired by osmosis from my culture. Rather, these four credos address an ethical realm that my culture has altogether ignored: human relations with other species, ecosystems, Earth, and the time-developmental process of the pageant of life. None of these credos casts an ultimate value in terms of its use in securing an even greater human good. Such human-centered ultimates are all too familiar,[20] and a restatement of them is not the purpose of this book. Human-centered valuation of nature (sometimes called *resourcism*) can be transmitted entirely by appealing to rationality and self-interest; you don't need religious conviction to value species, such as salmon, that provide important human services, or to value clean air and scenic terrain. But you do need something other than self-interest to spark concern for a snail darter.

Overall, the four credos reside entirely outside anthropocentric concerns. We value the pageant of life, the diversity of life, and so on because they are expressions of our creation story. These credos are admittedly anthropogenic (human-generated), but they are far from anthropocentric (human-centered). I have come to them by way of science only by reaching out and covering with my palm the bright sun of anthropocentric valuation, so overwhelming in my culture. Only by doing so have I a chance to see and admire the birds and insects and clouds aswirl in the surrounding sky.

The "way of science" scouted thus far in this book does not necessarily lead to these ultimates. Since the time of Darwin, a far more common interpretation of evolutionary cosmology has bred anthropocentric rather than ecocentric values. Interpretive flexibility and subjective inflection should, however, be no more fatal to the efficacy of this epic—however told—than it has been for the Bible. Similarly, one can hold an evolutionary cosmology and even judge biodiversity to be the derived ultimate value without caring a whit for the wild environments that have served as the cauldron of evolution for billions of years; rather, one can believe it is time for humans to take charge of cosmic evolution, thereby substituting the rapid response of genetic engineering for the slow and mindless wanderings of natural selection.

Nevertheless, the point here is that for me and a growing number of others, the cosmic story, centering on the earthly story, can indeed be rendered in shades of brilliant green—without departing a smidgen from the underlying science. There are, of course, fields of science not covered in this book that can also be used to nurture green values and ecoreligious perspectives. The disciplines highlighted here are nontrivial examples, and they are, moreover, the only aspects of science whose ecophilosophical extensions I can convey with knowledge and conviction. Those who derive a big part of their greenness from quantum physics, in contrast, have long had Fritjof Capra's writings as a source of inspiration. More recently, chaos theory and the complexity sciences have attracted their green proponents, too.

Four credos impart pluralistic richness rather than unific consistency in my bedrock beliefs. As will become evident later in this section, pluralism is pragmatically useful, too.[21] Pluralistic ethics offers an expanded toolkit for approaching the nuances of particular questions that emerge in the real world. Like the yin–yang binaries, one or another credo takes the lead in my psyche, depending on mood and situation. For emotional wholeness as well as practical use in formulating (or justifying) my opinions on a range of environmental issues, I need and want them all. Nevertheless, all four personal credos are grounded in the same underlying cosmology. There lies the ultimacy, but it is a cosmological not a value-assertive ultimacy.

Values to which I am predisposed can be made to emerge from the underlying cosmology. Recall from "A Conversation with Catalysts" in chapter 2 that the green morality shared by the six of us who told our personal stories actually preceded the development of personal cosmologies. Cosmology is thus not the only way one constructs a sense of meaning and value; in our cases, the experience of witnessing assaults on nature by the society in which we lived was sufficient to nurture the morality. But remember, too, the exhilaration Loyal Rue felt when he discovered the evolutionary epic: "My moral commitment finally had a foundation. All of a sudden, whoomp! It goes down like a taproot. Now it's grounded in the cosmos." More to the point, with respect to this cart-and-horse conundrum, all six of us are now taking actions that we hope will catalyze change, so that future generations will, in the normal course of growing up, be presented with joyful, science-based cosmologies that nurture green value systems—well before children are exposed to the sadnesses of human-caused extinction and environmental desecration. Our hopes are that an intertwined cosmology and ecocentric value system will be imparted by families, offered to children in bedtime stories, and experienced in whatever institutional or noninstitutional religious instruction and ritual a child may encounter.

Let's take a closer look at each of the four proposed credos.

The evolutionary epic is my creation story, and the pageant of life is evolution's great achievement. This statement embraces Ursula Goodenough's "credo of continuity," presented in chapter 2. We herald the time dimension of the entire pageant—including an overwhelming desire for this Great Dance of Life (borrowing from Dave Foreman[22]) to continue. Ours is thus a creation story of immense temporal scope and which is still very much in progress. It is through the study of living beings and their fossil forebears that the story of creation has been revealed and interpreted. Thanks to paleontologists, we can know and therefore honor the stories of those who came before: the ancestors, the pioneers, the heroes. As to the future, we take care to keep evolutionary possibilities open by respecting the processes—which even our best scientists understand but dimly. "The unknown evolutionary destinies of other life forms are to be respected," urges Gary Snyder.[23] Further, if we identify our own interests with those of the pageant (as we regularly do with our children and grandchildren), then much of the pang is removed from the recognition of personal mortality; as long as the pageant continues, "we" carry on. On the dark side, we take comfort in knowing that the pageant *will* continue with or without our own species aboard the ark—as it has been, is now, and forever shall be. We know too that, given sufficient time (perhaps ten million years), the biological impoverishment caused by a mass extinction of our own making will be repaired.[24] Knowledge of past extinctions and recoveries thus offers us a reasonable and comforting faith. "This too shall pass," the fossil record proclaims, but replenishment will take place at the pace of geologic time—not human, ursine, avian, or even piscine time.

The evolutionary epic is my creation story, and the diversity of life is evolution's great achievement. Here is a green version of the classic "principle of plenitude." This credo draws attention to our kin, our companions, in one particular slice of evolutionary time that we call the present. It is the pageant of life in the here and now. For those who hold to some version of this credo, the diversity of life is scripture. To behold another species with reverence is no less a religious act than to read the Bible in a pious frame of mind. To lose a species through apathy or avarice is no less tragic than to delete a passage from every Torah or to tear a page out of every copy of the Koran. We value the richness of biodiversity as a whole and therefore all the species and all the within-species genetic diversity that expresses it. As Gary Snyder reminds us, each and every species is "a pilgrim of four billion years of evolution."[25] Every lineage is thus to be honored for what Holmes Rolston calls its "storied achievement."[26] More, following Edward O. Wilson, some of us may judge an increase in diversity as the central progressive element that can be detected (and celebrated) in what is known about the pageant of life. Accordingly, this second credo may, for some, be derivative of the first. For others, it shines in its own right, in part from what we have come to feel by having experienced the richness and glorious particularity of other life forms. Overall, like the first credo, this is a value statement that expresses the deepest convictions of those among us who yearn for an end to the sixth great extinction.

The evolutionary epic is my creation story, and the richness and integrity of bioregions is evolution's great achievement. Here we turn to an ecosystem and landscape perspective that values not only life but relationships among the living as well as the contextual support that makes life possible—the particular watersheds, soils, topographies, and regional climates that have evoked, by evolutionary processes, the particular species native to place and the particular character of *this* place as a whole. Bioregional *richness* encompasses the biodiversity concern of the previous credo—but biodiversity is situated within a richness of diverse habitats. Bioregional *integrity*, in turn, speaks to the relationships forged among the living and the nonliving, from which emerge capacities for ecological health and vigor: self-maintenance, resilience, and (over the long term) innovation. An emphasis on the integrity of ecosystems calls up the "land ethic" espoused by Aldo Leopold a half-century ago and brilliantly defended by Baird Callicott today.[27] Leopold will forever have a presence in my own Upper Gila bioregion. It was here that he had the transformative experience of watching "a fierce green fire" fade in the eyes of a dying she-wolf that he himself had shot; it was here that his ecocentric ethics were born; and it was here—and because of his efforts—that protection of landscapes in and for their wilderness character got its start. His legacy is the Gila Wilderness, and just to the east lies an area that was protected after his death: the Aldo Leopold Wilderness.

The evolutionary epic is my creation story, and this self-renewing living planet is evolution's great achievement. Terrestrial bioregions plus the vast realm of Neptune (perhaps parsed by cetaceans into bioregional provinces, as well), and all sharing the same sheltering canopy of cloud and vapor, combine as Gaia. This is the biosphere, stunningly able to self-maintain and self-renew. Such capacities probably emerged when life was still exclusively microbial. Gaia is thus far older than any living species, any ecosystem. The diversity of life is the living tissue of planetary geophysiology. The evolutionary pageant of earthly life takes place strictly within the spatial confines of the biosphere. In a way, therefore, this credo embraces all the others. But, observing the reverse side of the Möbius strip, one can embed the planetary credo, in turn, within the overarching pageant of life—which, after all, gave birth to and forever shall maintain the self-renewing talents of this world. This is a credo surely felt, if not so articulated, by the originator of Gaia theory. Jim Lovelock has written, "I find myself looking on the Earth itself as a place for worship, with all life as its congregation. For me this is reason enough for doing everything that is in my power to sustain a healthy planet."[28]

Applying the Credos

Overall, these four credos mean that, for me, the pageant of life is sacred, the diversity of life is sacred, bioregions are sacred, Gaia is sacred. Let's try them out. Here is a real-world question of environmental ethics that has been on my mind of late, and about which I have testified in a public hearing and written comments to a federal agency. It concerns the Mexican gray wolf, the southernmost and rarest subspecies of gray wolf. A few Mexican gray wolves may still be alive in remote mountains of northern Mexico, but in the American Southwest all have been exterminated. A captive-breeding program was begun in the United States in 1978 from a founder population of seven wolves. The captive-bred wolves now number about 140, dispersed among two dozen zoos and captive-breeding facilities around the country. The U.S. Fish and Wildlife Service has proposed to reintroduce the Mexican gray wolf into the Apache and Gila National Forests (within the greater Gila bioregion). The question: Should the reintroduction take place? In this imaginary dialogue, DA is devil's advocate, who holds to a highly developed set of ethics—but these are exclusively human-centered.

CB: I believe the Mexican gray wolf should be reintroduced into the wild.

DA: Why?

CB: Because the Mexican gray wolf is an expression of biodiversity, and biodiversity is sacred.

DA: Why is biodiversity sacred?

CB: Biodiversity is sacred because it is my religion to believe so.[29] *The evolutionary epic is my creation story, and the diversity of life is evolution's great achievement.*

DA: Well, it's not my religion to believe so. The Earth exists for the material support of humans. Species, such as deer and elk, that have recreational value for humans or that are crucial for ecosystem maintenance are important to sustain, but for everything else, human needs should take precedence. As I see it, the controversy is between the ranchers and hunters on the one side wanting to keep their livestock and wild deer safe from predation versus you environmentalists on the other side wanting to hear wolves howl in the wilderness.

CB: You've got my side completely wrong, I'm afraid, and that's precisely why I feel the need to declare the depth of my beliefs. For too long, conservationists have used the vocabulary of human interests to press our concerns. The debates have thus been framed in terms of my aesthetics, or my recreational interests, versus your economic interests. But many people who support wolf reintroduction do not expect ever to hear a wolf howl in the wild. Rather, those of us who feel this way—theists as well as nontheists—regard the natural world as an expression of something sacred. The issue is thus deeper than aesthetics, or recreation, or economics. To knowingly cause the extinction of another evolved or created species is a sacrilege. Public lands managers are beginning to realize that environmental controversies are no longer a matter of arbitrating among competing human interests. It's a religious issue for a growing segment of the population.

DA: Be that as it may, you should recognize that the Mexican gray wolf subspecies has been sustained and even bolstered during the past two decades of captive breeding. Failing to reintroduce the wolf into the wild will not, therefore, mean its extinction. So why can't the authorities just keep the lineage going in its captive-breeding program?

CB: Captive breeding is a stopgap measure. Ultimately, a species—especially a highly social species with a rich behavioral repertoire—loses its identity, its natural way of being, unless its members are provided once again with a chance to live as organisms in a rich and amenable environment—preferably their native habitat.

DA: Well, maybe this all should take place in Mexico; after all, there may still be a few wild Mexican wolves there.

CB: Wherever the reintroduction may first occur, the Mexican wolves should ultimately also repopulate the greater Gila bioregion.

DA: Why?

CB: The greater Gila bioregion is one of the very few regions in the American Southwest that contains a large core of mostly uninhabited wild country, where wolves have a chance to live unencumbered by human cultural influences that evolution has not equipped them to cope with. Before the Mexican gray wolf was exterminated from the United States, it was a native and integral member of my bioregion. My bioregion is diminished in its absence.

DA: Why does it matter if your bioregion is diminished by one species? The ecosystem doesn't absolutely need the wolf. We are now using, and can continue to rely on, sport hunting as a management tool, to take the place of the wolf in controlling deer and elk.

CB: I believe that hunting and other forms of management in our country's yet-impoverished patchwork of designated wilderness areas should not be in lieu of natural processes that would otherwise be available. These deliberate human interactions with the ecosystem should mesh with and be subsidiary to spontaneous natural processes.[30]

DA: Why?

CB: For one thing, history has shown that we are often wrong when we think we understand the workings of nature well enough to mimic or substitute our own for natural processes. Something unanticipated happens. Those of us who follow the way of science are humbled by the vastness of human (and scientific) ignorance. More fundamentally, my vision is of a world in which significant realms remain in a natural state. This is because *the evolutionary epic is my creation story, and the richness and integrity of bioregions is evolution's great achievement.*

DA: I don't agree with you, but I see no point in arguing on that level. Nevertheless, why not at least put an end to this expensive captive-breeding program? Before reintroduction can take place, those wolves will have to be *trained* to be wild and to avoid humans and their livestock. So why not just haul in some surplus wolves from British Columbia instead?

CB: The wolves in British Columbia are a subspecies that would be non-native here; they may not be as well suited as the Mexican subspecies for this climate and landscape.

DA: Well, as I recall, you now have a thriving population of elk in your bioregion—the Roosevelt subspecies. This non-native subspecies was brought in from Yellowstone National Park, as your own native subspecies, the Merriam elk, was hunted into extinction before anybody cared enough to save it.

CB: True, the Roosevelt elk seems to have fit in well as far as we can tell, but the loss of the subspecies was an irreplaceable loss of genetic diversity and therefore an irreplaceable loss of evolutionary potential. A subspecies is an early step toward speciation.

DA: Why should evolutionary potential and speciation be of any concern to us?

CB: Because evolution is in Earth's future as well as the past. *The evolutionary epic is my creation story, and the pageant of life is evolution's great achievement.*

DA: But again, maybe the best way to secure the gene pool that remains is *not* by subjecting the wolves to the dog-eat-dog world of life in the wild.

CB: Securing the gene pool by captive breeding is only a stopgap measure. Ultimately the potential for evolution of this lineage must be restored. All organisms are organisms-in-environments; biodiversity as a whole is the pageant in process; and wild lands and waters are the arenas of evolution, where we too have a chance to refresh our souls, our ancestral memories, in the kinds of wild conditions that brought forth all the various cultures.

DA: Oh no, I sense that damnable refrain on its way again!

CB: *The evolutionary epic is my creation story, and the pageant of life, the diversity of life, and the richness and integrity of bioregions are evolution's great achievements.*

DA: Let's get practical here. Assume, for the moment, that all your lofty ideals are taken as a given. Does it make sense to be spending so much of limited funds on just one endangered lineage?—and a mere subspecies at that!

CB: I'll grant that objection as one well worth considering. But these agonizing choices I prefer to discuss with people who share my ultimate beliefs. Nevertheless, there are two things I can say to you. First, the work of paleontologists digging up bones and dung from the end of the Pleistocene suggests that we humans may have already driven to extinction two-thirds of the large mammal genera that filled the Americas as recently as 12,000 years ago. In my view, therefore, the large mammals that do remain should receive extra attention. Second, consider that it is easier for humans everywhere to identify with charismatic megafauna than with the less spectacular animal species or plants. Reintroduction of the Mexican wolf is not therefore just for the benefit of the wolf, biodiversity, evolution, and this bioregion; it will also help move people to care and to regard reintroductions as opportunities for joyful celebration in a world too often overwhelmed with bad news.

DA: Not everybody is planning to joyfully celebrate, you know. There are ranchers who feel they will be carrying the full economic burden, because they may suffer losses among their livestock that now graze on public lands—including much of the core, designated wilderness. How do you respond to these human problems?

CB: The ranchers have a valid concern, but it is not insurmountable in my view, because the government has pledged to outfit with radio collars released wolves (and their pups born in the wild) and to relocate or even kill any wolves that establish territories on private land or that prey upon privately owned livestock grazing on public lands. The wolves are in no way being released into the "wild." Their lives will be watched and their actions constrained. The wolves will be given freedom to roam on our terms—not theirs. Ultimately, I hope our societal values will shift to where we move toward phasing out cattle and sheep grazing in core wilderness areas; there must be *somewhere* that wild animals will be granted the chance to live among themselves, unencumbered by the economic intrusions and resource priorities of populous industrial societies. Sadly, that is not where our nation's values seem to be yet. Therefore, the concerns of the ranchers who have a tradition of grazing livestock on public lands in the wolf reintroduction area must be dealt with responsibly. I would, in fact, feel uncomfortable advocating wolf reintroduction if there were not a way for *me too* to share in the financial risk—especially because our own neighborhood rancher, whom I respect, does indeed graze most of her livestock on public lands in the Gila Wilderness. Fortunately, the Defenders of Wildlife organization has set up a private fund to compensate ranchers for wolf kills, and I have contributed to that fund.

DA: Well, I'm not convinced. I don't share your core values. I still think wolf reintroduction is crazy and pointless. But I'll grant that your position is supportable on the basis of your core beliefs. Nevertheless, it seems to me that although you environmentalists may espouse ideas that are not overtly misanthropic, you really don't see much value in the human species.

CB: Good point. Failing to celebrate our own worth can contribute to low species self-esteem, with all the problems that come from that. So how about another credo? *The evolutionary epic is my creation story, and the emergence and enrichment of meaning is evolution's great achievement.*

DA: I don't see *Homo sapiens* in there.

CB: Sorry; I should have clarified. We can perhaps consider ourselves the meaning-makers of the cosmos. More than any other species we know of, humans have the capacity to find meaning in, and thus care for, just about everything. We can care about the fate of more than ourselves and our families. We can care about other people in our bioregion—be they friends or foes of biocentric ideals. We can care about people on the other side of the planet. We can care about the Mexican gray wolf and other endangered species, too. We can even care about evolution continuing into the future—heck, we're probably the only species that even knows about evolution. It would be a tragedy for the cosmos to lose that capacity, by losing us. You, DA, may not find biodiversity or the pageant of life meaningful, but there is a chance that your great-grandchildren will. You should not deny them the chance to live out their lives in as biologically rich—and potentially meaningful—a biosphere as you and I have been blessed to live within. And so I say to you, *the evolutionary epic is my creation story, and the emergence and enrichment of meaning is evolution's great achievement.*

At this point CB, grateful for the excellent service performed by DA, bids her interlocutor farewell. The meaning-making credo will be explored in detail in the next section, but first, there are several points I'd like to highlight from this imaginary dialogue.

Academic philosophers who work in the field of environmental ethics often invoke *intrinsic value*—positing that a group of organisms, such as the Mexican gray wolf subspecies, has value in itself, value for its own sake. This is an important argument, because whatever possesses intrinsic value need not be *instrumental* for anything else in order to merit a continuing place on this planet. The holder of intrinsic value is an end in itself, not simply a means to an end that is deemed the greater good. Intrinsic value effectively means that an appeal on behalf of a lineage need not refer to a presumed human economic, aesthetic, or even spiritual benefit. A species or subspecies with intrinsic value need not have a role to play in erosion control, in maintaining nature's balance, or even in the continued well-being of other species dependent on it. Note that the credos proposed here do place a kind of ultimate value, or value-in-itself, in five things: the pageant of life, the diversity of life, rich and integral bioregions, a self-renewing living planet, and meaning. According to this way of thinking, it doesn't matter whether the Mexican gray wolf is judged to have intrinsic value. That point never comes up in the dialogue because each and every species and subspecies is taken by CB to be an *expression* of something that is indeed afforded the status of an ultimate value—a thing (the diversity of life) or a process (the pageant of life). Also, every species is an expression of its native bioregion.

Similarly, there is no "rights talk" going on here. The notion of species rights is tricky for a number of reasons. Baird Callicott warns, "To extend rights to wild animals would be in effect to domesticate them. . . . Instead of imposing artificial legalities, rights, and so on on nature, we might take the opposite course and accept and affirm natural biological laws, principles, and limitations in the human personal and social spheres."[31] In other words, humans and human cultures are embedded in the natural world; we should not attempt to embed the natural world in our cultural concept of rights. Arguing the same end, proponents of transpersonal ecology (more widely known as deep ecology) urge that we cultivate ecological consciousness rather than legislate rights for wild creatures and lineages.[32] By so doing, we encourage a wider scope of caring and more expansive standards of virtue. Only then will ecologically right action come naturally. Beautiful acts spring from the heart; dutiful acts demand codes of right behavior.

From the ecofeminist side, too, come objections to invocations of species rights. These arguments are sometimes grounded in Carol Gilligan's landmark study of gender differences, which found males leaning toward "a code of justice" and females toward "a code of caring."[33] Rights have an important place in the former worldview, personally felt bonds of commitment in the latter. These bonds are not imposed from outside by societal code; rather, they well up naturally within humans who see themselves not as isolated individuals protected from a hostile world by an edifice of rights but as interdependent members of a community sustained by a network of reciprocity.

Conservation Priorities

The proposed credos may provide a framework for approaching the lamentable need for triage. Where should the limited money that is available now go? Who among the crowds of imperiled taxa will receive our attention? For whom are we most willing to sacrifice a bit of our own wealth and comfort?

To begin, by placing an ultimate value in the **pageant of life**, we surely must argue for an all-out effort to prevent all deeply rooted *"living fossils"* from vanishing. After tens, even hundreds, of millions of years of continuity, surviving episodes of widespread extinction, the Old Ones are simply too precious to lose because of our own avarice or apathy. The coelacanth fish, lungfish, tuatara reptile, horseshoe crab, remipedia crustacean, tailed frog, lingula brachiopod, nautilus mollusk, ginkgo tree—these would receive our utmost devotion. All are the last surviving members of distinctive (and in some cases, once-mighty) lineages more than a hundred million years old. Consider the ginkgo tree, which co-inhabited the Jurassic with dinosaurs all over the world but which is now found wild in only a few mountain valleys of China and mostly in deliberate plantings. It is the only living member of its genus, its family—indeed, an entire order (Ginkgoales). From a taxonomic standpoint, loss of the single species of ginkgo alive today would be equivalent to losing all 610 species of pine, fir, spruce, cypress, and cone-bearing kin that compose the order Coniferales.

Imperiled lineages that may not be deeply rooted on the tree of life but nevertheless are highly *disparate*—that is, separated from the most closely related lineage by a great taxonomic distance—would receive special attention as well.[34] The giant panda, for example, would fare well under this criterion; the lineage is only about ten million years old, yet its closest relative is so remote that there is disagreement about whether it belongs in the bear family, the raccoon family, or a family all its own. Similarly, the hoatzin of the Amazon rainforest would be considered a very valuable expression of the pageant of life; it is the only "ruminant" bird (having a fermentative foregut), the only bird whose young bear a claw at the bend of each wing, and the only living member of its taxonomic family. A very young lineage that expresses *a burst of evolutionary creativity* would also be precious, as a group. Here the cichlid fishes of Africa's huge but isolated rift lakes come to mind. In Lake Victoria more than three hundred species evolved from a single species of parent stock within just a few tens of thousands of years. Sadly, the stocking of Lake Victoria with a commercially valuable predatory perch from the Nile River pushed more than half of those species into extinction within the past two decades.[35]

We would also take extra care with those lineages that *express pinnacles of sensory achievement, have unique adaptations, or are singular in some other way* (such as the largest or smallest mammal, bird, fish, tree, and so on). The Komodo dragon, largest lizard in the world, and the Chinese water dragon, largest amphibian in the world, would rank high under this criterion. So would an east African caecilian, *Scolecomorphus kirkii*; this wormlike creature is the only vertebrate with highly mobile, protrusible eyes. Naked mole rats for their antlike social allegiance unknown to other vertebrates; leafcutter ants for their well-tended fungal gardens; many other ants for their husbandry skills in tending and "milking" aphids for a sugary secretion: all these are stunning, and therefore precious, examples of evolutionary achievement. So, too, are the bog-meadow plants (such as the Venus flytrap) that feed like an animal, and the mouthless flatworm (*Convoluta roscoffensis*) that feeds like a plant, dependent on photosynthetic green algae embedded within its translucent tissues. Even the overly ubiquitous cockroaches that have adapted to human dwellings would be valued by this criterion, as they are the fastest runners on six legs. Cicadas, with their synchronized life cycles as long as seventeen years, win the prize for the insect group with the longest juvenile phase.

We would value, too, *elaborate mutualistic relationships* among species, such as the orchid and wasp pairings that have produced stunning flowers in the quest for perfect pollination. And phenomena, not just taxa, would be valued—notably, the great, seasonal migrations of African wildebeest, Arctic caribou, monarch butterflies, California elephant seals, Pacific humpback whales, various species of salmon, and numerous birds. A single cave in Romania would merit the highest conservation efforts; not only does it contain thirty-three species of invertebrates found nowhere else, but it is also the only terrestrial community known to have a food chain entirely dependent on chemosynthetic bacteria at the base.[36]

All these judgments of exceptionalism and achievement are a bit squishier than the taxonomic criteria used to recognize living fossils and taxonomically disparate recent species, but there is nevertheless some possibility for consensus. Finally, what about *future evolution*? Ay, there's the rub! As Christian de Duve warns, "What will be recognized tomorrow as a fork organism is a mere terminal twig on the tree of life today."[37] Then too, because our concern with the pageant of life is not simply to preserve a record of a story that has ended but also to ensure that the story will continue, preserving tokens of species in the form of "minimum viable populations" inhabiting small islands of habitat is only a stopgap measure. We aim, ultimately, to restore to vibrancy what Dave Foreman calls "the arena for evolution"—that means not merely conserving but expanding and regrowing wildlands.[38] Ours is thus a creation story that is far from over.

In terms of the pageant-of-life ethic, how would the Mexican gray wolf fare? The subspecies has zero standing as a living fossil; moreover, if the subspecies vanishes, the species still survives. It is, however, the southernmost subspecies of the gray wolf, which makes it a singularity of sorts and probably the most heat-adapted of the wolves, which may prove important in a warming world. Overall, however, the Mexican gray wolf does not fare well in triage when the pageant of life is the sole basis for judgment.

How next does the **diversity credo** impinge on protection priorities? As advocated by Norman Myers and others, conservation efforts should be focused on *"hot spots" of biodiversity*.[39] Notable hot spots of global concern are the islands of Hawaii, Indonesia, and Madagascar. Continental hot spots in North America include Mexico, generally, for rodents and bats and reptiles, the southeastern United States for freshwater mussels and snails, southern Appalachia for amphibians, and coastal California for just about everything. Protection of *umbrella species* in each of these areas—that is, species that require lots of acreage and relatively pristine conditions—effectively protects many additional species with less demanding requirements. As in the pageant protection standard, taxonomic *disparity* would become decisive in circumstances in which particular species rather than concentrations of imperiled species must be prioritized. But the beauty of the diversity criterion is that, in the main, it supports first and foremost an *egalitarian* regard for species. All species great and small are expressions of the diversity of life.

By egalitarian standards of diversity protection, the Mexican gray wolf would be just one of thousands of imperiled subspecies crying out for help, and any full species of snail or flowering plant would surge ahead of it. Moreover, it is likely that many more species of endangered invertebrates and plants could be protected if the money now being spent on captive breeding for this one subspecies of vertebrate were diverted to other conservation efforts. Pragmatically, however, the Mexican gray wolf might jump far ahead in the queue. For one, you can't store animals of any kind in a seed bank. In contrast, an arctic lupine seed, recovered from permafrost ten thousand years old, germinated and grew into a healthy plant.[40] No giant ground sloth will ever emerge from an alchemy of fossil bones, hair, and dung. Animals that depend on parents to teach them the ways of the world and to inculcate social skills cannot be fully preserved as frozen embryos or DNA. Jurassic Park simply will not work for these species, because cultural achievement is not in the genes. As mentioned by CB, captive breeding is only a short-term measure for holding on to a highly social species of skilled hunters. Finally, the Mexican gray wolf shines as an umbrella species—and one, moreover, with large-mammal charisma that may be useful in expanding popular concern for endangered species programs in general.

A **bioregional perspective** for triage places importance, in part, on the functional roles played by species in their native habitats. An all-out effort must be made to preserve *keystone species*, for example, and other species judged critical for ecological integrity. A species might not be imperiled from a global perspective, but it may be very rare near the fringes of its range and may therefore prompt bioregionalists to take special efforts to retain its presence in their home areas. Some who adopt a bioregional perspective may be motivated to work to protect *endangered species* that absolutely depend on that region; others may judge some such species to be too functionally trivial and inconspicuous to merit attention. Rather, funding priorities may go to *restoration work* to shore up degraded hillsides, river banks, and salt marshes—perhaps, but not necessarily, offering benefits to endangered species. Restoration efforts aimed at watershed protection may prevent soil or shoreline erosion that would otherwise take centuries to heal. An old growth forest might take five hundred years to regenerate from scratch. That's a long time in human terms, but it must be kept in mind that extinction is forever. From a bioregional perspective *endangered ecosystems*, perhaps more than endangered species, would capture the greatest commitment.

Bioregional loyalty may be a good; bioregional insularity surely is not. To encourage concern for bioregions beyond one's own, it might be helpful to apply the "veil of ignorance" exercise developed by John Rawls in his *Theory of Justice*.[41] In the lottery of life, we might have been born into or have made our home in *any* bioregion. How then do we set priorities among the bioregions?

Within its native bioregions, the Mexican gray wolf merits a privileged position for protection efforts. Top predators are, by definition, critical species. Among area environmentalists, reintroduction of wolves should feel like the return of a lost loved one. Only the ghosts of wolves remain in the greater Gila bioregion. Protecting wild remnant populations of a species is one thing; reintroducing an organism (and a charismatic one, to boot) that was exterminated by bounty hunters a half-century ago should be cause for celebration. It is a sign of our spiritual progress, our widening circle of concern. We thus demonstrate psychological maturation beyond the toddler stage of species self-interest. No more a self-centered two-year-old, our species begins to realize that the world does not exist exclusively to satisfy our own desires. There are others with interests, too.

Finally, from a **gaian perspective**, priority goes to the protection of *vast photosynthetic landscapes and seascapes*—notably forests, coastal wetlands, and the plankton-rich "meadows of the sea"[42]—all crucial to the maintenance of planetary climate and chemistry. Particular species are usually of little concern from this perspective. Gaia needs forests; it doesn't matter whether the trees are maples or oaks, firs or pines. More fundamentally, Gaia needs photosynthesizers; it may not matter whether chlorophyll molecules occupy grasses or deciduous trees, golden diatoms or coccolithophores. The priority is scale, not particularities. *Redundancy* of certain kinds is crucial over the long term, however. As Lynn Margulis advises, "Diversity, in the gaian context of regulation of atmospheric gases, is not a luxury. Millions of differently functioning types of organisms are necessary for the maintenance of flexibility and responsiveness in the face of inevitable astronomical and geological perturbations on the planet."[43]

Especially vital from a gaian perspective are those taxa whose *geophysiological functions are unique*—that is, not duplicated by others. For example, of the more than fifteen thousand species of legume plants, those that are known to produce root nodules in symbiotic associations with nitrogen-fixing bacteria all depend on three or four genera of Rhizobia bacteria. Very few—possibly only one—species of ericoid fungus associates with a host of bog-dwelling flowering plants. Without essential symbionts, these plants (spanning three taxonomic families) would be unable to obtain minerals and nitrogen from highly acidic soils. Moving to the marine realm, free-floating algae (especially coccolithophores) may be the major producers of dimethyl sulfide,

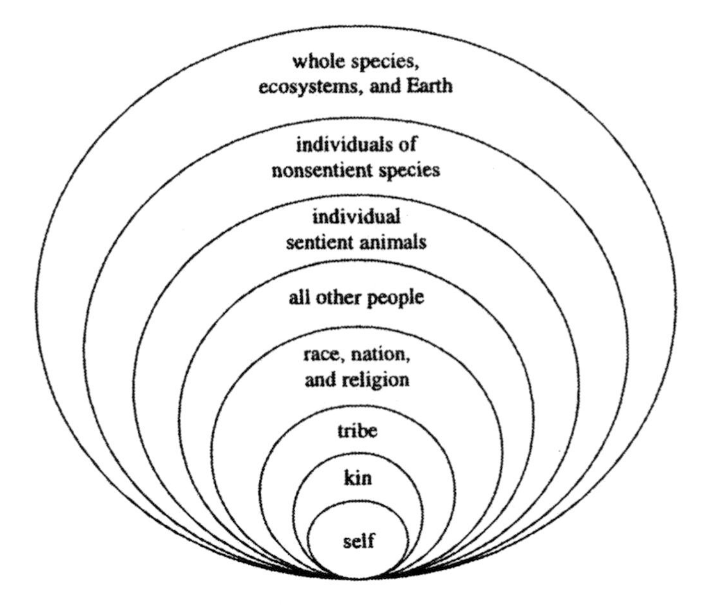

EXPANDING THE CIRCLE. Adapted from Reed F. Noss, 1992, "Issues of Scale in Conservation Biology," in P. L. Fiedler and Subodh K. Jain, eds., *Conservation Biology* (New York: Chapman & Hall).

which diffuses into the atmosphere to play a key role in marine cloud formation. Overall, the species that may be most important for the health of the biosphere are, at present, far from endangered. The question, rather, is a matter of scale. There must be sufficient forest, sufficient wetlands. More, the system must not be overwhelmed by human inputs—notably chlorofluorocarbons and carbon dioxide vented to the atmosphere. Global warming is a threat not only to coastal cities and established agricultural zones; it may devastate many fellow travelers of the Cenozoic, especially those terrestrial plants and animals now confined to islands of wilderness surrounded by civilization and therefore incapable of edging toward higher latitudes to track the warming. *Humility* is the watchword in a gaian perspective. Even with the best science, we cannot know for certain how much human-caused change can be absorbed, how much extinction—and of what—can be tolerated by the gaian system in its present mode.

How does the Mexican gray wolf fare from a gaian perspective? It doesn't even register on the scale of concern. (That was why it was the only credo that CB never invoked in the dialogue with Devil's Advocate.) Overall, Gaia would probably continue to function with barely a hiccup even if all vertebrates vanished—although it is possible that Earth's climate might shift in consequence. Indeed, to persist, Gaia may need nothing more than bacteria, which effectively run the global metabolism. Consider that for the first half of the pageant of life, Earth was an entirely bacterial world. An exclusively bacterial manifestation of Gaia would surely foster different environmental conditions from those of today, but there is little doubt that the system itself would continue to thrive.

In summary, with just four credos that place ultimate value in the pageant of life, the diversity of life, bioregions, and Gaia, there will be agonizingly difficult choices to make in setting conservation priorities. A pluralistic set of credos that share a cosmology but not an ultimate, single value does not lend itself to a cookbook approach to reconciling internal differences. Answers are gloriously gray, and slippery slopes keep us on our toes. Intuition, feeling, the concerns and recommendations of others, and all the nuances of context will come into play. A full and fallible human being—not a logician—participates in value decisions. Yet however strained and assailable the judgment, the underlying cosmology stands firm.

All this is complicated by the fact that Earth is now home to more than five billion human beings who are busy making even more. Many, far too many, of those billions lack decent prospects to secure their vital material needs, much less personal fulfillment. And already this single species has appropriated to its own ends nearly half of Earth's photosynthetic output on land.[44] As the population grows, the consequences of human fallibility—intellectual and ethical—become increasingly burdensome on ourselves and on other species. The environment can compensate for fewer and fewer of our mistakes. How, then, do we think of the human?

From the standpoint of the evolutionary epic, this is a difficult question. Humans have been around for only a blip of geologic time. Evolution was in full swing for eons before we came on the scene and would still present a marvelous basis for story even had *Homo sapiens* failed to evolve to tell it. Similarly, we may not do well in the evolutionary lottery that determines species longevity—especially if we continue our assault on the carrying capacity of the environment on which we depend. This means that the evolutionary epic may well sweep into the future without us. The pageant of life, the diversity of life, bioregions, and Gaia: All these earthly features or processes have long and distinguished pedigrees that run in the billions of years. *Homo sapiens* is simply not in that class. Nevertheless, we can devise a credo that makes us an important *expression* of an ultimate value with a very long pedigree. We can do this, moreover, without treading the old and dangerous path of anthropocentrism.

The evolutionary epic is my creation story, and the emergence and enrichment of meaning is evolution's great achievement. This emergence and enrichment of meaning is the subject of the next section.

Addendum: In March 1997, U.S. Secretary of the Interior Bruce Babbitt approved a plan for reintroducing the Mexican gray wolf into the greater Gila bioregion.

Meaning-Makers

"Aunt Connie, what are black flies for?" My nine-year-old niece Halsey poses this excellent question while we are enjoying a several-hundred-acre stretch of cranberry bog, tamarack swamp, and oak-maple forest in central Michigan that was once owned by her grandfather and has passed on to her father. It is an insect-free autumn day, and Halsey knows that something called "black flies" is the reason she has never visited this landscape during the summer.

"Black flies are food for the migrant songbirds who fly up from the south to raise their babies here during the summer." Then I ask a question in return, "What are humans for?"

Halsey looks startled, so I begin, "When our ancestors first came down out of the trees, our job was to eat fruit and walk around a lot so that we would spread the seeds of fruit in our poop. That way the seeds could sprout in rich soil and grow into new bushes and trees. Most of us don't go out and poop on the ground anymore, so what are we good for today?"

"To hunt the deer to keep them from overpopulating," she answers brightly.

"Good!" I encourage her, "Anything else?"

What *are* humans good for? We know all too well today what we are bad for. "A cancer on the planet" is a common description of our self-made ecological role ("humanpox," writes Dave Foreman[45]). We are, as Neil Everndon maintains, an exotic species out of control, "suspended in ignorance, capable of material existence but not of community commitment."[46] We are the perpetrators of the sixth great extinction. We have breached the limits nature placed upon our early primate heritage. Clothes, shelter, and fire on demand now allow us to live far beyond the tropical savanna to which we are native. Food we no longer just find, but plant and irrigate and husband and process and store and transport vast distances. To slip past the once-formidable executor of population standards, we medicate and we vaccinate. As to predators, we have long since eliminated most of them. But plundering the planet is a self-imposed job description—not our fate. To do less harm is thus a crucial goal for our species, but if that is *all* we can envision for ourselves, then the best we might do would be self-annihilation. Bryan Appleyard accordingly paints environmentalism as "a religion of catastrophe. We can only undo the harm we have done; we can aspire to nothing higher."[47]

Low species self-esteem is the specter haunting Halsey's generation. And with that come additional perils for the Earth—the self-centered bravado of compensatory puffery, the retreat into live-for-the-moment consumerism, or the lassitude born of hopelessness and despair. We long for a positive role to play while we work to lighten our impress. One positive role is to compensate for the misdeeds of past generations. Rather than simply doing less harm, we actively work to heal the landscape, restore vitality to polluted waters and eroded soils, and roll back the extravagant fringes of civilization, thus inviting wildness to reclaim lands from the merely rural. As with the Mexican gray wolf, we can return species to bioregions from which they were exterminated.

Another possibility—though one not widely endorsed within the environmental community—is that, while we are busy doing less harm and making amends for our misdeeds and those of our forebears, we can turn our most sophisticated technologies toward service of the Earth Community. We can begin an all-out effort to detect comets and asteroids whose orbits might someday intercept that of Earth, and we can work toward developing safe methods to nudge the intruders aside. (I am writing this at a time when comet Hale-Bopp is a benign splendor in the predawn sky.) We would thus become Gaia's immune system, preventing future mass extinctions so long as our technological kind abides on Earth. Philosophically, this is a tenuous position. Where would mammals be today had *Velociraptor* or some other brilliant lineage of dinosaur been able to prevent the meteor impact that put an end to the Mesozoic era? What is the natural course of evolution anyway? To protect the planet from a cataclysmic impact is, of course, a self-serving thing for us to do. But we would also be acting in behalf of many of the large creatures we care about, fellow travelers of the Cenozoic. Ultimately, envisioning ourselves as the planetary immune system is a position one takes, or does not take, on faith. Science and technology may give our society the option, but the role is one we will choose or choose not to pursue.

At root, underlying any sense of a cosmic or an earthly role for humans is a grounding image of how we relate to other beings, the planet, and the cosmos. Several philosophers have of late scrutinized the range of environmental worldviews and value systems and have detected three distinct types of species self-image.[48] One self-image has contributed to our long history of ecological malfeasance. This is the *isolated self*. We are observers of, not participants in, the cosmos. Other names for this viewpoint are the Cartesian self (after Descartes), the atomic self, and the existential self. Although such terms make this standpoint sound unremittingly bad, it does have an up side. Notably, the isolated self adopted by many scientists in their work (under the banner of objectivity) has made possible a great deal of scientific insight that we can now use to build a species self-image that is far from isolationist.

A second possibility is a *relational self*. "In the beginning is the relation," wrote Martin Buber more than half a century ago in his book *I and Thou*.[49] In this view of the self, other beings and other aspects of the cosmos are no longer distinctly "other"; they are relations. Each is a "thou," not an "it," according to Buber. A relational self establishes bonds of kinship with other species. Diane Ackerman expressed this idea in chapter 3: "I believe we should regard all life forms with dignity, respect, affectionate curiosity, and the kind of protectiveness family members feel for one another." Holmes Rolston calls this perspective "the entwined self."[50] Deane Curtin puts forward the Buddhist philosopher Dōgen, who lived in Japan eight hundred years ago, as the intellectual precursor of a relational self-image.[51] A relational sense of self is the image generally espoused by ecofeminists, too, but for them such a philosophical stance needs no famous personage to give it form and credibility. Rather, a relational self-image is the sense of self that comes naturally to those engaged in child rearing.

The third possibility is an *expanded self*. Each of us is more than a skin-bounded body with skull-encapsulated brain. We are intimately connected with all our relations in our bioregion, upon this planet. This spatial extension of self has been called "eco-self" by Joanna Macy [52] (in contrast to ego-self). Holmes Rolston beautifully portrays this spatially expanded sense of self: "The human vascular system includes arteries, veins, rivers, oceans, and air currents. Cleaning a dump is not different in kind from filling a tooth. The self metabolically, if metaphorically, interpenetrates the ecosystem. The world is my body."[53]

Another sense of expanded self is temporal. I, for example, am not forty-five years old; I am pushing four billion. My evolutionary self is the very pageant of life, which shall carry on even when this body and mind known as Connie Barlow reach the end of their limited term of existence. Here is Holmes Rolston again: "Nature treats any particular individual with a momentary life, but life is a propagating wave over time. Located in individuals, value is consigned to a stream." My favorite exposition of death comes from philosopher Loyal Rue, who, like me, embraces the evolutionary epic. In a work in progress, Rue extends into the realm of meaning the new scientific understanding of the relation of death and sexually reproducing (hence, complex) organisms. The segregation of the germ line from the soma (body tissue) line of cells made sexual reproduction possible—and death inevitable. We now know that the former simply could not exist without the latter. Yet it is an expanded sense of self, brought by way of science, that allows Rue to accept death not with equanimity, but with outright enthusiasm:

> In the wisdom of this scheme it becomes difficult to view death in a negative sense. The inevitability of my death is now beheld as a necessary condition of the life I have. A mere entrance fee, to be paid at the exit. If there were no death, there would be no soma line. And without a soma line, there would be no possibility of an embodied person—no memories, no loves, no wonder or wisdom, no longing or learning. These are among the splendors of the body, and for these we must die. To the extent that I cherish my life, therefore, I have reason to be profoundly grateful for my death.
>
> How then shall I think about death? With gratitude, and as the occasions decide. When I have occasion to mourn the death of others, I will try to absorb the loss in what I have gained from them. I will try to understand my grief as a measure of my gratitude. And when I have occasion to consider the fact of my own death, I will attempt to think large. I will try to see that a soma-centered story of the self is a small and impoverished view, and that the life within me was first quickened among the primordial organisms appearing on Earth nearly four billion years ago. I will affirm that all lives, no less my own, are instruments of life itself. And by these measures I will submerge the absurdity of death in gratitude for the wonder and wisdom of life.

An expanded self-image in both space and time is espoused by the originator of deep ecology, Arne Naess,[54] who borrowed the concept of a greater self from Spinoza. Recall the remark made by Stephan Harding, a proponent of deep ecology, which was recounted at the end of the previous chapter. Harding said that what he was seeking in his scientific studies was not knowledge for the sake of knowledge, but rather knowledge that would nurture a wider and deeper "identification" with the natural world.

Ecofeminists, however, sometimes criticize this expanded self-image as a hegemonic takeover of the other by extension of one's own ego. Deep ecology advocates disagree. According to Warwick Fox, "What identification should not be taken to mean is *identity*—that I literally *am* that tree over there, for example."[55] Nevertheless, the expanded self as identity with the other is a shared facet of mystical experience in virtually all religions; it forms the basis for what Aldous Huxley called "the perennial philosophy."[56] This kind of expanded self is at the core of the Hindu credo "That thou art."

A way to reconcile the relational self and the expanded self is to cultivate a sense of *embedded self*.[57] Contemplation of Gaia is a good way to do this. Recall Tyler Volk's comment in the previous chapter, "Learning about the science of Gaia makes me feel a lot more connected with the Earth. When I go for a walk in the woods now or just look up at the clouds, I feel like I'm inside a giant metabolism." Volk is not that tree over there, but tree and he are intimately related: They are both cells within the planetary body of Gaia. What tree breathes out, he breathes in. From a gaian perspective, the only self-sufficient whole is the self-renewing biosphere. Everything embedded within the biosphere is both whole and part—whole unto itself and part to something greater. Arthur Koestler coined the word *holon* for this fusion of part and whole, and Tyler Volk has applied the concept of holon to his understanding of Gaia.[58]

Another way to reconcile the tension between a relational self and an expanded self is perhaps by way of a *communal self*, which again is a kind of holon. As an individual, I relate to other individuals in the Earth Community, but I also identify with the well-being of the whole such that a transgression of my community is a transgression of my self. A communal sense of self draws from the same root that Thomas Berry uses as his grounding metaphor: *communion*. "The universe is a communion of subjects," avers Berry, "not a collection of objects."

Brian Swimme reminds us that even Gaia is not a self-sufficient whole. The surface biosphere depends utterly on the shower of photons offered by the sun. Ultimately, the cosmos is the only whole that—as far as science can tell—is completely self-contained, self-sufficient.

What should be the root metaphor of our society? What root metaphor can nurture a turn toward an ecological consciousness? Ian Barbour, a philosopher and historian of religion, traces the shift in root metaphors underpinning Western societies.[59] In medieval times the root metaphor was *kingdom*. Not surprisingly, religious interpretations issuing from that era would speak of the "kingdom of God." The root metaphor changed at the onset of the scientific era. Isaac Newton and those who followed him depicted the cosmos not as a kingdom but as a *machine*. At first, like all machines, the cosmos had a creator; much later the creator was dropped, but the machine imagery lingered. Finally, today, Barbour detects a new root metaphor emerging: *community*. Again, Thomas Berry offers *communion* because it means all that *community* means and more; *communion* carries a sacramental quality as well.

In the process of writing this book, I grew an attachment for a rather different root metaphor: *conversation*. This metaphor is evident in the very structure of this work. Conversation is, after all, a community in process. Conversation binds individuals into communities. Similarly, ecosystems, bioregions, and Gaia emerge and are sustained by a kind of physical and chemical conversation among the living and between the living and the nonliving. Evolution is an extension of those same conversations through time. *Conversation* suggests spontaneity, mutual creativity. It has no pre-established destination, yet we can count on something interesting developing. When we give our full attention to the present moment, in effect losing ourselves in conversation, an immense and delightful journey unfolds—effortlessly. A conversation is thus a kind of miracle. So too is the immense journey of the evolutionary epic.

Perhaps we should have multiple metaphors, with *community* the taproot metaphor and *communion* and *conversation* and others the branching roots.

Thus equipped with a survey of the various images of self (isolated, relational, expanded, embedded, communal) and with musings on root metaphor, we can revisit the question *What good are humans?*

The days are over when good-hearted, thinking people could believe that Earth and the universe were created for human benefit and pleasure. By way of science, we have come to accept the humbling truth that the cosmos carried on quite well for billions of years without us and will continue to do so when we are gone. By way of science, we know that Earth, this solar system, even this galaxy are infinitesimally tiny particles in the grand sweep of the universe. Moreover, Earth, at the moment, is decidedly worse off because of us. And yet, and yet, we do have gifts to offer. We can enrich the Earth Community and the cosmos with our presence. Further, if we limit our numbers and our toll on the landscape so that the planet can heal, far more humans will have a chance to enjoy and enrich the cosmos in the future—because our species will have bettered its prospects for longevity. To reverse the world's population trend is therefore not only a biocentric but an anthropocentric goal. The true misanthropes are those who argue for business as usual.[60]

So, then, what inspiring image can we carry of ourselves as we begin the hard task of changing course? How can our presence enrich the cosmos? What are our gifts?

Edward O. Wilson (chapter 2) suggested that we view our own species' contribution as that of "life become conscious of itself."

Holmes Rolston proposes a slightly different self-image: "We are the love of life become conscious of itself." Also, "In humans, an evolutionary ecosystem becomes conscious of itself."[61] Half a century earlier, Julian Huxley urged us to consider ourselves as "evolution become conscious of itself."[62] Thomas Berry and Brian Swimme envision our kind as the very universe become conscious of itself.

Ten years after our species brought forth the first image of Earth from space, James Lovelock wrote, "The evolution of *Homo sapiens*, with his technological inventiveness and his increasingly subtle communications network, has vastly increased Gaia's range of perception. She is now through us awake and aware of herself. She has seen the reflection of her fair face through the eyes of astronauts and the television cameras of orbiting spacecraft. Our sensations of wonder and pleasure, our capacity for conscious thought and speculation, our restless curiosity and drive are hers to share."[63] Similarly, Joseph Campbell has said, "We are the consciousness of the Earth. These are the eyes of the Earth. And this is the voice of the Earth."[64] By all these standards, therefore, we are life, evolution, love of life, Gaia—even the cosmos—become conscious of themselves. All are *expanded self* images.

Arne Naess, Annie Dillard, Stephan Harding, Gary Snyder, Ed Dobb, Elizabeth Johnson, Thomas Berry, and Brian Swimme all, to some degree, share a vision that moves consciousness to the next step: We are not just to know but also to *express* what we know. Naess envisions the human as "the conscious joyful appreciator" of this planet. This, from Gary Snyder: "If we are here for any good purpose at all (other than collating texts, running rivers, and learning the stars), I suspect it is to entertain the rest of nature." Annie Dillard advises, "You were made and set here to give voice to this, your own astonishment." Stephan Harding, in conversation, offered this lovely idea: Our species' role is "to give poetic expression to the world." My friend and fellow science writer Ed Dobb has written eloquently in a similar vein for *Harper's* magazine: "As language-using organisms, we participate in the evolution of the universe most fruitfully through interpretation. We come to grips with the world by drawing pictures, telling stories, conversing. These acts are our special contribution to existence. . . . We are aware of the stars only because we have evolved a corresponding interior space—the domain of mind and language—in which stars can be reflected and in whose absence the cosmos would exist but cosmology would not." Dobb concludes, "It is our immense good fortune and grave responsibility to sing the songs of the cosmos." "We are the cantors of the universe," proclaims theologian Elizabeth Johnson.[65]

To sing the songs of the cosmos is precisely how Thomas Berry and now also Brian Swimme envision our species' role, our self-image. In all their writings, this message is central: We are celebrants of the universe story. But we are not just cheerleaders for the universe. We *are* the universe celebrating itself. Here the expanded self and joyful expression merge.

"Good!" I say to Halsey. "What else are humans for, besides keeping deer from overpopulating?"

Halsey is stumped, so I continue. "Just think! We are the only species that can write poetry, paint pictures, tell stories."

One day Halsey will surely read or hear someone say that humans are a cancer on the planet. She will witness ecological destruction and learn about species that were alive when her aunt was born but are no more. I hope these realizations will spark an activism in her. But I hope she will remember that our kind has gifts to share—and she does too. I hope she will regard herself as life roused into awe-struck wonder of immensely diverse ways of being. I hope she will feel in her own flesh the sense of Gaia awakened and aware. I hope, too, that she will find herself cheering on the becoming of life: the past, the present, and the future pageant. I hope she will continue the conversation as an informed, caring, and passionate celebrant of the universe story. Thus I hope she will come to see that meaning-making is the very meaning of life for meaning-makers such as we.

Living Dangerously

"The trouble with following the path you are on," Dick Holland cautions me, "is that if you hang your star on a new idea in science, and it turns out not to be true, then you're left without a star."

Heinrich (Dick) Holland is a professor at Harvard University, and he represents (and works on the cutting edge of) the mainstream of geochemistry. He wrote the textbooks that everyone still uses. And he thinks the star I'm orbiting is a particularly dangerous choice. Holland is a staunch and outspoken critic of Gaia theory. "Life responds; it does not regulate," he insists. The stunning persistence of life on this planet is due not to biospheric self-regulation but to the adaptability of life and "the relative dullness of earth history."[66]

Dick Holland, critic of Gaia, is not, however, a critic of Jim Lovelock—in fact, they have become great friends. "Jim and I go fairly far back. He and Lynn [Margulis] used to come here, and he would ask me how I thought the system worked. I would give him an impassioned one- or two-hour lecture on the way I thought it worked, and he would thank me profusely and say that really clarified a lot of things—but, you know, it never made any difference."

After a bout of shared laughter, Holland continues, "And then when he wrote the first book and I dealt with it in my atmospheres and oceans book, concluding that the Gaia hypothesis was charming but not correct, I thought that was it. But no, no, Jim being Jim, he thought that was okay. He kept coming back for some sort of understanding from my point of view. And, you know, everyone loves that man. He has a magnificent way of attracting people to himself. At the Scientists on Gaia meeting, I was sitting next to him at the banquet table and he leaned over and said, 'I'm so relieved. I thought people were just going to tear me into little bits.' And I said, 'Jim, nobody's going to tear you into little bits; we're all just too fond of you.' " Dick Holland, with his gentle voice, ever-present smile, and talent for disagreeing without dismembering, could very well be describing himself.

Holland may be one of Gaia's greatest critics, but he plans to join the just-hatching Geophysiological Society nonetheless. "As long as that same gang gets together, I'll keep coming to the meetings." He is, moreover, green to the core. We all sensed this at the Gaia conference in Oxford, when, during a wide-ranging group discussion of science and values, he launched into a soliloquy that brought me (and many others) to the brink of tears. I asked him for a short retake of the story.

"When I go jogging on weekends, Esther, my granddaughter, likes to come along on her bicycle. One day we were passing Mystic Lake, near where we live, and decided to stop. We went down to the water's edge, but partygoers had left a horrendous mess. I said, 'Well, if we're going to stick our toes in the water, we've got to make sure there's no glass in there.' One thing led to another. We did a lot of cleaning up. Maybe half an hour. That's a lot for an eight-year-old. After it was all done, she said very quietly, 'I think we've helped the Earth.' "

"She didn't say, 'I think we've helped Mystic Lake,' " I add. "She put it in terms of the Earth."

"She's very special. It came out of her heart. It didn't come from me. And it didn't come, as far as I can tell, from other members of the family. The kids will be told to clean up, but there's not that sense of doing it because you're helping something, you're helping the Earth. It's because your Mom and your Dad told you to do it. And then, for Martin Luther King Day in school, she wrote something like, 'I have a dream that in communities there will be many gardens, and people will take care of the environment and not litter.' You get enough kids like that and I think the world will be okay."

Dick Holland teaches an undergraduate course at Harvard on the environment. One of the texts he assigns is his own, *Living Dangerously: The Earth, Its Resources, and the Environment.*[67] The book includes everything I would want to know about earth system science, and more. The title indicates that it is packaged for relevance to what should become everyone's concerns. It is a tool for nurturing citizens knowledgeable and thoughtful about our planet's (and hence our own) biggest problems. Holland may be green, but, as it turns out, it is not because of his science.

"What is it about the Earth system," I ask, "the cycling and so forth, or the changes through time, that you find most inspiring? What gives you a sense of awe or wonder or magnificence?"

After a long pause, he responds, "Well, I feel a little differently about it. I feel more like somebody who opens up the gearbox of a car and gets his hand in there and says, 'Ah! Now I know how this thing works!' That to me is the excitement, the excitement of the chase, of trying to understand how this big, complicated system actually works. In some cases the answers have been forthcoming fairly readily, and in others, the oxygen business, for example, they've been very slow in coming because it turns out that the interplay of living organisms, dead organisms, and the inorganic processes that shape, for instance, the availability of nutrients is by no means trivial. But it's been great fun trying to understand how these things fit together." Another silence, and he turns toward the window, "It's not the science. I draw much more inspiration from looking out there and seeing fruit trees in bloom."

Holland continues, "At least to my way of thinking, it doesn't really make any difference whether the biosphere does this or that.

We have responsibilities to the world as a whole, one of which is to find out what's going on. But whether or not Jim's first position was right, or his current position, or something that is still hidden in the mists of time, I don't think is relevant to the question of responsibility, our responsibility as stewards of the Earth."

Stewardship is a notion that I do not fully empathize with—except as intended by Paul Mankiewicz in his felicitous expression, "hands-off stewardship." But Holland feels strongly about this issue and brings me up short.

"Stewardship has nothing to do with philosophy or theology," he admonishes. "It's just knowing that we are changing the face of the planet in a major way. We're the dominant animal species. We've learned how to control so many of the things that used to control us—disease, for example. So, like it or not, the role of caretaker or exploiter—whichever way you want to look at it—is thrust upon us.

"Does Hagar the Horrible mean anything to you?" Holland continues, "He's a comic strip character, a Norseman. He sits there with his bristly beard, a slight paunch, and his beer, and he says, 'I've got mine!' " Gestures accompany the punchline, and then Holland's face grows serious, "Stewardship is the alternative. Anything else is short-sighted. Now, this doesn't mean that we have answers to all the problems that come up. But we must do our best."

So, this is Dick Holland's view of where the science comes in. It helps us act, and act responsibly; but it is not the science that makes us *want* to act responsibly. The impetus must come from somewhere else, quite possibly an experience in childhood that happened well before we opened our first science book—something like Esther's experience with her grandfather at Mystic Lake.

Dick Holland recalls his own story, but he can't articulate a causal explanation or an Esther-type epiphany to explain the roots of his green values. In the early 1940s, while still in high school (a Christian boarding school in New York), he became a fundamentalist Christian.

"As a fundamentalist Christian," he explains, "I had a very clear view of our responsibility to the Earth, but within the framework of our responsibility to God and God's dealings with us."

"Are you still a fundamentalist?" I ask.

"No. But it took a long time to extricate myself from it. At Columbia [where he received his Ph.D. in geochemistry] I was president of Intervarsity Christian Fellowship, and then I went to teach at Princeton. For a number of years thereafter I would have counted myself a fundamentalist. But gradually the thing started to split. The split came between looking at the Earth using geology as the book and then having to square this with the notion of God as an active participant in every part of everyone's life. So my religious view deteriorated. What happened is sort of fascinating in retrospect. I came to realize that the religious superstructure had been built on top of and as an integrating structure for my own personal and deeply held view of the world. When my faith deteriorated I discovered that I didn't really need it. What remained was something like what the view of the Earth from space offered so many people decades later, and which is now such an important grounding for all of us. We have a sense of responsibility for the Earth. I have found that while this view is certainly promoted by the Christian worldview, it does not require a Christian or any other religious basis. It is something that flows out of my own feeling for ethics."

"So does the ethic now mesh with the science for you?" I ask.

"You can't derive ethics from science. One can be a superb scientist and still be a real bastard. In my way of thinking, there has to be a clean split between the questions of how the world really works and how we should respond ethically and morally as a planetary species."

Tracking back to Holland's personal story, I ask, "Once your religious faith unraveled, did you still go to church?"

"Yes," he says, "Fact is, I still taught Sunday school into the sixties at the First Presbyterian Church in Princeton. But I was considered backslidden."

"I don't associate Presbyterian with fundamentalist," I say.

"Oh yes," Holland responds, "There are fundamentalists in most of the denominations, although there are some major exceptions. You will not find fundamentalists among the Unitarians, for example."

"Right, and that's what I am. You'll find everything else in our denomination—including paganism—but not fundamentalism!"

We both laugh and then Holland continues. "My son Matthew, a historian, is now the administrator for the Unitarian-Universalist church in Weston [Massachusetts]. A couple of years ago he asked me, 'Just exactly where do you stand on this issue of God?' I answered, 'Well, it's very difficult.' 'Try,' he suggested. And I said, 'Well, intellectually I don't see it anymore; emotionally I'm still very tied to it.' "

Dick Holland, as a young fundamentalist, had unwittingly embarked on what turned out to be a dangerous course by pursuing a research career in the earth sciences. He knows what it is like to have one's faith undermined. Because he has also lived through some wrenching paradigm shifts in geology, he cautions me against building faith and pinning ethics on any notion derived from science. He thus cautions against living dangerously.

John Maynard Smith offers the same counsel, but for very different reasons.

"I think it is dangerous to conflate myths with scientific theories," he tells me. "And it is immensely dangerous to want our scientific theories to become myths or to underpin our myths. The danger of that is you will finish up with bad science because you'll want it to be good morality."

I am speaking with John Maynard Smith during a scientific conference on Gaia at Oxford University. He has made substantial contributions to evolutionary theory, notably through his invention of mathematical models to explore the interworkings of chance and selection. His most recent book explores how innovations in the way information is transmitted in the living world underpin what he calls "the major transitions in evolution."[68] Maynard Smith, one of the world's most respected evolutionary biologists, was invited to the Gaia meeting by Jim Lovelock to serve as a friendly critic.

"If you want science to give you some kind of mythical reassurance," Maynard Smith cautions, "then you may start believing bad science. And I think that's dangerous. That's why I'm worried about Gaia, to be honest. If we do in any way treat it as a myth, by which I mean an account of the world which is a justification for green politics, if you like, then the danger is we'll distort the science in order to get the answer we want."

He continues, "I'm very fond of Jim Lovelock, and I think he's done a very useful thing in many ways. But I do think it was unfortunate he decided to use the name of a goddess. However rational he may be about his theory, the name is bound to mislead people, to cause people to believe things about the Earth because they want to, and not because they're true. I don't know how far it's true that the Earth might be a self-regulating system, but I would like to think about the problem completely free of any mythical overtones."

Maynard Smith speaks from experience on the matter of mixing science and myth. In his youth during the 1930s, he became a Marxist because Marxism offered the strongest reaction against the fascism then building across the English Channel. Later, however, he learned that Stalin had used his power as leader of the Soviet Union to promote an academically discredited but ideologically attractive theory of genetics through its chief proponent, T. D. Lysenko. As a result, the discipline was destroyed in the Soviet Union. Worse, dissident scientists who would have been hailed in the West for their research were sent to camps in Siberia, and thus often to their deaths. "It was a frightening experience," Maynard Smith recalls.

"What about today's extensions of evolutionary biology into the realm of meaning?" I ask. "What about E. O. Wilson's claim that we can regard the story of life as an epic? What about those of us who want to promote the evolutionary epic as our creation story? Is this dangerous?"

"Wilson is not the first to suggest this, you know."

"True," I respond, "but his predecessors are no longer speaking out. They're dead."

"I remember Dobzhansky," Maynard Smith muses. (Theodosius Dobzhansky was a leading geneticist of mid century, who also wrote books and essays promoting the evolutionary story as a scientific form of religion.) "I once said to Dobie, 'How could you write that rubbish approving Teilhard de Chardin?' And he said, 'John, if people have got to have a religion, it had better be an evolutionary religion.'

"I didn't find that convincing," he continues. "But I may have changed my mind somewhat over the years. I do think that human beings need myths. They need stories with some kind of moral meaning, which place them in the universe. They don't necessarily have to believe in the factual truth of those myths. You can be moved by *Hamlet* or *Macbeth* without believing those events actually happened. The factual truth is not essential to the meaning derived."

I press him on this point. "So," I say, "on the one hand you agree there is an ethical, even a religious, need. But you don't see any problem in maintaining the old myths. You would keep the old creation myths and simply treat them in a more whimsical fashion, trying to get rid of the fundamentalist notions and treating those myths metaphorically."

"In general I think that's true," he responds.

I continue, "So you don't have an answer for what humankind needs to do with its myths to make them more ecologically attentive and reverential. You would just prefer that science not be used in this endeavor. You don't think we ought to mine science and evolution in particular for a new global myth."

"No, I don't ," he says. "Obviously science is highly relevant to the issue because it's important to know, for example, whether global warming is occurring or not. That is a scientific question. But, no, I don't want to derive myths from evolutionary biology—though I'm sure I have used the argument that for something that has taken three thousand million years to evolve, it would be a pity to crash it down by our sheer mindlessness. I do think that. And I find biodiversity moving. Ever since I can remember, I have been a passionate naturalist. I like watching birds; I like growing plants; I like keeping beetles—I still do. I became a biologist because I love nature. I do not love nature because I became a biologist."

He pauses, then continues. "I guess I have some confusion in my mind. I grant that we need myths—human beings do, I do. I don't live without values. I don't live without things I mind about. To an important extent the things I mind about were influenced by stories I was told as a child and the things I've heard since, including nonscientific things. Yes, I do think we need myths, but I'm uncomfortable with the deliberate construction of myths, because I'm afraid of their distorting effect on the science."[69]

For those of us who wish to follow "the way of science" in cultivating our deepest, even religious, worldviews, how do we respond to these two dangers? How do we individually address the personal danger pointed out by Dick Holland? How do we collectively deal with the danger to scientific inquiry enunciated by John Maynard Smith? Are the personal and societal risks worth courting?

Not long after my conversation with Dick Holland, I settled on a personal response to the danger of "hanging my star on science." If being religious means to have faith in something for which there is no assurance, then I have faith in my own creativity to respond to the inevitable changes in science in ways that allow me to feel like a natural being at home in a remarkable universe. I can therefore live quite comfortably with a book of revelation that comes with the promise of errata sheets. As to Maynard Smith's warning of a societal danger, I was clueless. I decided to ask friends for their thoughts and to help me hone my own through conversation.

"Julian, how do you respond to the dangers pointed out by Dick Holland and John Maynard Smith?" I have called upon Julian Huxley, whose books have kept him very much alive in my mind, even though he died in 1975. I first made his acquaintance ten years ago, when he introduced me to the evolutionary epic and thereby changed my life.

"May I remind you, Connie, that I lived my life and wrote my books in exuberant denial of the specter that seems to haunt Maynard Smith and which has made him exceedingly wary of any commerce between two of the highest realms of human creation: science and religion. You see, in my mind, for us to allow our religious groundings and cosmologies to limp along farther and farther behind our moving picture of the world—which is, today, bestowed by science—is to court disaster. Such a profound schism is not sustainable over the long term. Wouldn't you agree?"

Before I have a chance to answer, he continues, "As to the matter of personal wholeness and the point made by Professor Holland, I think I have something you might find helpful." He fades for a moment and then returns with my 1994 anthology in hand, which contains extracts from his own writings and that of many other philosophically inclined biologists.[70] "Ah yes, here is a pertinent paragraph or two." He pauses for a moment and looks up, catching me square in the eye. "You seem to have overlooked a number of passages of utmost importance from my writings, you know, but on the whole, you did rather well."

"Thanks, Julian," I respond. "Which paragraphs provide an answer to Dick Holland?"

He scrutinizes the text. "Hmmm. The language about 'man' and so forth is dated, I see, but nevertheless the idea still holds, I think you will agree." Huxley clears his throat and begins his recitation, "In all spiritual matters we should expect steady change and improvement as man accumulates experience and perfects his mental tools. As matters stand today, we have the cleavage between orthodox religion which has hitched its wagon—the sense of the sacred— not to a star, but to a traditional theology, and a large body of educated people who, rejecting the theology, are forced to stand outside religion too."

"How prescient of you!" I interrupt. "You take the star image presented by Dick Holland and turn it around. It is because we have not yet fully hitched our wagon to the stars that so many of us committed to a scientific worldview have thrown out the baby with the bathwater."

"Exactly," he concurs, "Far too many of those who follow the path of science have stifled their innate religious capacities when all they really need do is cleanse themselves of the ancient orthodoxies. Surely you know that every one of my books was written to counteract the regrettable trend toward spiritual oblivion. Religion is simply too precious to allow the theists to make it exclusively their own."

He returns to the recitation, "In art it is a triumph if a Beethoven or a Titian finds new ways of building beauty; in science it is acclaimed a triumph if an old universally accepted theory is dethroned to make way for one more comprehensive; but in the religious sphere, owing largely to this pernicious view that religion is the result of supernatural revelation and embodies god-given and therefore complete or absolute truth, the reverse is the case, and change, even progressive change, is by the great body of religiously minded people looked upon as a defeat. Whereas once it is realized that religious truth is the product of human mind and therefore as incomplete as scientific truth, as partial as artistic expression, the proof or even the suggestion of inadequacy would be welcomed as a means to arriving at a fuller truth and an expression more complete."

Loyal Rue responds to my query with an e-mail: "So the cosmology changes. Is that a big deal? Not really. It just gives me good reason to update my religion. Where does it say that religions can't be updated just as routinely as science? Indeed, one might say this: I hitch my religion to science because that ensures that my religion will continue to grow. Hey, imagine that: a religious orientation that grows and adapts to changing conditions!"

I follow up the e-mail with a phone call to Loyal. I tell him his sentiment is exactly that of Julian Huxley, and then read him Julian's statement. Halfway through Loyal erupts, "Right! Right!" I end the reading with, "Now you see why I fell in love with this guy?"

"Yes." Loyal continues, "Well of course religion has to change. We have to be able to say it is not only incomplete sometimes but *wrong*. By God, we had that one wrong! I mean, science says that. If you're going to tie your religion to science, you've got to be able to say that."

"Science as a community is expected to admit error," I add, "but the scientist who invented the now-wrong theory is not expected to. It's the community of science that does it; the individuals can be as stick-in-the-mud as they wish. You wait for the old scientists to die and then a new theory can take over. And that's the beauty of it. Humans don't have to be divine gods in order to do science and to create from it a self-renewing, science-based religion."

Loyal continues, "We've misunderstood religion all these years. We've assumed that religion is unchanging, authoritative. Because it came from God, how could it be wrong? But now, with science, we know that truth is always tentative. And now truth is no longer received; it's earned."

"I like that: Truth is earned."

He continues, "These unearned truths, where somebody comes up with an insight because they've got a cramp or something and it makes them stay up all night and then they catch a spark of an idea which feels to them like a revelation from the gods: That's important; truths are always important. But those truths are unearned because they come through one subjective flash of insight. Rather, when those insights are then put to the test of scientific method and scrutiny and discussion, then an insight can be honed. Only then does it become an earned truth—and one earned by the community as a whole."

"Okay," I say, "let's assume you and Julian are right—that we got religion wrong all these years, but maybe as a result we've now evolved to *need* unchanging authority in our cosmologies. Maybe our Stone Age psyches can't bear a truth that is only tentative. After all, why do we call religious traditions *traditions*? Doesn't that tell us something?"

"Anybody who lives in this world," Loyal counters, "even long before Heraclitus, had to have been aware of flux. Change as well as stability is the way of the world."

"Maybe it has to do with the level of change we're faced with," I offer. "For example, most Christians would probably not be bothered if they were told that the Jesus who died on the cross maybe had darker skin than they had seen in their Sunday school picture books. Similarly, when I first learned about Gaia, I was impressed with the long-term stability of the biosphere. Later I had to adjust my thinking to find grandeur, too, in the huge changes in Earth's climate that may in fact have taken place over the eons. But that was a level of change I could cope with—though, I admit, not without some difficulty at first."

"Similarly," offers Loyal, "if what is deeply embedded in most people is some sort of notion that this is God's world, then you let them hold on to that. You ask them, instead, 'Where did you get the idea that God isn't continually repainting this canvas?' It's a work in progress, not a finished thing."

"So you can tell the epic and still let them keep their God," I surmise.

"Exactly," he says. "You don't go out there and say, 'Hey, you've got your thinking all wrong, so we're going to do a complete brain transplant.' Evolution never does that. Evolution says, 'Okay, let's see what you've got here and how we can build onto it. If this system has a big problem, let's see if we can do an add-on, and maybe that old system will eventually atrophy.' "

Our conversation continues and then shifts to the problem raised by John Maynard Smith: the danger to science itself if people begin to call upon it widely for their religious orientation.

Loyal responds, "I think Maynard Smith is looking for a kind of ivory tower that science just isn't going to have. Science *will* be politicized; it is a human institution. It will be vulnerable. All knowledge systems are vulnerable to human beings manipulating them. But the wonderful thing about science is that it has built into it some procedural checks and balances: you know, the squinty-eyed nervous anxiety about authority. It has built into it these purgatives that will release it from political bondage. There are self-correcting features in science that will always try to loosen the grip of politics on this form of knowledge. But to assume that science can be apolitical and out there beyond human manipulation is unrealistic. And, in fact, that would be a terrible thing because it would make science irrelevant."

A few days later, I am on the phone with Ursula Goodenough. "So there's a concern that this particular movement we're involved with will trigger political control over the scientific questions asked," she begins. "Well, that's not a concern I share. If that were in fact to happen way downstream from here, I would worry about that when the time comes. I certainly wouldn't see it as a reason not to do what we're doing now. More to the point, I think there already is huge control over the kind of scientific questions being asked. That's already happening in many aspects of science."

"What about the other danger?" I ask. "What about the personal danger of hanging your star on science?"

Ursula responds immediately. "Evolutionary theory, it seems to me, is no more likely to go away than atomic theory. It's not like this is some fringy kind of idea. The data supporting evolutionary theory are so deep, so robust, and come from so many completely different lines of inquiry, that as far as I'm concerned, it has the status of atomic theory. In fact, I made this point just a few days ago. I was leading a workshop on science and religion for college teachers, sponsored by the Templeton Foundation. The first day my topic was evolution. One teacher remarked that she told her students that evolution was a theory and that scientific theories are good only until the next one comes around. So I asked her, 'What do you think about atomic theory? Do you think this table is made up of atoms or not?' And she said, 'Well, yes, I think it is.' And I said, 'As far as I'm concerned, evolutionary theory is just as solid. I think the reason you have trouble with it is you have trouble with some of its implications. So let's explore those.' Which we did.

"Fundamentally, Connie, I don't think a good scientific theory—one that's based on empirical observation—ever gets disproved. It just gets amplified and deepened and incorporated into something broader. For example, the whole addition of complexity theory doesn't disprove evolutionary theory at all. Rather, it amplifies our understanding of how the phenotypes might be generated."

"So it has to do with the level of theory," I say. "If you hang your star on the general concept of a naturally occurring evolution through time, it's a pretty good bet that your star is going to stay hanging."

Ursula agrees. "Richard Dawkins makes this point about evolutionary theory in one of his books: that there really aren't any competing theories. When you think about it, there just isn't any other way the sequence of life could happen."

"It's either a natural process by evolution or a supernatural process by divine fiat," I concur. "So you and I are so convinced evolutionary theory is going to hold up, at least in the main—the same way as you say atomic theory has—that we are willing to make the leap into a kind of religious orientation. We're not going to be disappointed. If to be religious is to have faith in something that you have no absolute total proof of, then you and I have faith that we can hang our star on evolution and build a religion around it for ourselves because we have faith it is always going to be there for us."

Ursula disagrees. "I'm leery about that word. Faith is a different thing altogether. I mean, do you have faith that the sun is going to rise tomorrow morning, or are you pretty damn sure?"

"I guess I'm pretty damn sure. So does that mean you and I really don't have faith in anything?"

"I would say that the faith comes much later," she explains.

"There's a leap that comes somewhere down the pike. For me the faith is a hope, an expectation, that human nature will transcend its fear and greed."

Last, I decided to solicit a response to the "living dangerously" questions from a gaian perspective. One of the geophysiology researchers you met in the previous chapter is not even a phone call away. I live with him. The conversation began one morning with this question: Is there any discovery that might possibly happen in geophysiology that would traumatize your worldview?

"That's a really good question." Tyler Volk pauses, then continues. "I guess I would say no. If it turns out to be just lucky that life has persisted on Earth—that life has not in fact played a pretty significant role in its own success—then that's going to be mind-blowing too because it will make our situation very special. Whether life is a fluke and we are alone in the universe or whether life gets going and then takes charge of its own environment—either way is awesome."

He continues, "In terms of Earth, I'm convinced that 'just lucky I guess' versus Gaia as a self-regulating superorganism, the Great Womb, is too harsh a dichotomy. The truth is probably somewhere in between. Clearly, the atmosphere is a biological construct. But whether the atmosphere is a biological construct for life's good—that's a big question. Clearly, too, the soil is a biological construct. And parts of the ocean are biological constructs. How those things work together as a system over time is a question that is still wide open, which is why I am so fascinated with working on it. What we discover is going to affect how we see the evolutionary process and ourselves."

I probe a little further, "In the conversation with the other F_1's, you mentioned that knowledge of geophysiology has given you some real spiritual goods—that when you walk outside, you feel as if you're inside a giant metabolism. Is there anything that could be discovered about Gaia that would destroy that feeling for you? How vulnerable is that feeling to the way of science? And if it gets scuttled, what could take its place that would be of equal spiritual value to you?"

"Well, I don't see how it could be scuttled," he responds, "as this feeling is based on —I hesitate to say it—but it is based on immortal facts. We absolutely know that oxygen is given off by plants and algae, and we well know that the oxygen reservoir has been net created by their burial. We absolutely know that animals and fungi and other respirers take in what plants and algae give off. And we absolutely know that the atmosphere serves as a great circulating fluid connecting all the organisms around the world."

"So those things, those cycles, just aren't going to change," I conclude.

"Think about it," he continues. "We were learning that stuff in third grade."

"Well, I wasn't."

"Okay, but a long time ago," he insists. "Somehow, though, it wasn't put in the context of, 'Isn't this incredible, kids?' Or, 'Let's go out and do a little dance in the courtyard and breathe with the planet.' "

We both laugh; then he continues. "To me the cycling is a source of contemplative mystery, maybe the way Jesus is to those in my family who still follow Catholicism. It is a source to spin off into bigger thoughts."

"So, the way of science, yes, it is open to change," I surmise. "But you nevertheless have a faith—or, as Ursula puts it, there are certain facts you are pretty damn sure of. These facts and others you find very attractive. And you know that either they are unlikely to change or, if they do change, they are going to be supplanted by something that is even more awesome."

"Yes. I'd like to say 'No, the fundamental pairing of photosynthesis and respiration during the carbon cycle won't change.' But I realize that position is a kind of faith."

"A justifiable faith, too," I add. "The postmodern critics of science say that because we are subjective and living inside the biosphere, the cosmos, we can never really know anything for sure. And I say, well, no. It's really a matter of scale. I know that the rock these cliffs are made of is volcanic. That's not going to change."

"These were hard-won facts in the recent few hundred years," Tyler concurs.

"Whereas if you get into more intricate causal systems," I continue, "the facts and theories are far more questionable."

"Right: like how the planet operates."

"That's the problem with the extreme postmodernists," I continue. "They don't make a distinction between the foundations that we can be pretty damn sure of and the far more speculative theories that might be tossed out tomorrow. But then, one of the beauties of the entrenchment of paradigms is that by the time we are forced to give up an old way of thinking, a maverick scientist or group will have developed a whole new way of understanding the world. The old paradigm simply hangs on until a new one boots it out. So when we have to give up an old foundation, we're not left without a god. There's a new god, ready-made, just waiting to be acknowledged. And it's likely to be at least as enticing a way to view the universe. So I think it's reasonable to have faith. Because even if the things you view as rock-solid turn out not to be, there's going to be something else right there for you. You're never going to be left out in the void."

"And if one accepts the hard-won revolutions, both big and small, as being parts of the story," Tyler adds, "then one doesn't just arbitrarily run to the next god that looks interesting—like in the sixties, we all knew somebody who embraced Buddhism for six months then all of a sudden became a Jesus freak. But with the science we can say, 'Whoa, look what we now know we were wrong about!' Say, for example, if we find out that the mass extinctions were not internally generated by Earth or the biosphere but all came from comet or asteroid strikes. If that shift in scientific thinking does take place, then our view of evolution will be hugely different—but it will be just as, if not more, exciting. And we can get excited, too, about having been wrong. Wow, isn't that great: we were wrong!"

"Oh!" I respond. "So instead of the change being something we're embarrassed about and we go turn to another religion, we're proud and thrilled that there never is an end to science. The story is never over."

"Yes, that's the beauty of it," Tyler continues. "These great events are not only in the past. They're still to come."

"It's like thinking about having been able to live in the time of Jesus," I muse, "maybe even to have known the guy. Well, we are living in the time of Jesus, from the way-of-science standpoint. Somebody you or I know might well be the equivalent of Jesus, in that they may utterly change the world of science and therefore all the personal varieties of religious feelings drawn out of the science. A hundred years from now, somebody we know might be considered as Darwin is now."

Tyler continues, "If we really have within us this biological need to internalize a rather permanent story of what the world is about, then not having a story, or being confronted with the changing story of science, could trigger a personal crisis. What if, however, our story becomes the story of how stories change . . ."

"Then we can get satisfaction and a sense of security and pride in ourselves out of that kind of story, too," I finish his thought.

"So even though we can't ground ourselves in an immortal story anymore," he offers, "the immortality can be had in the story of how we make stories, of how we find stories through science."

"Which is the way of science! That's the story behind the story," I announce. "So your story of the changing story is basically an equal celebration of the universe and celebration of the human mind discovering how to know about the universe."

"Oooh. That sounds good," Tyler responds. "If we need an immortal story, say, based on our evolutionary psychology, then the most immortal part of our story is not the facts about the world but our current semi-immortal way of finding out those facts: It's a celebration of the scientific process."

"Another thing I find attractive about that idea," I say, "which is also an attraction for me in the evolutionary epic, is that it's a continuing story. It didn't just happen a long time ago and then that's it. It's happening right now and it's going to happen in the future, and we have a responsibility for that future. It's not just something that happened back with Copernicus and Francis Bacon. It's an ongoing story; we're living in the midst of the evolutionary epic, and we're living in the midst of the immortal story of science."

"You could say that we each have to be a microcosm of that story," adds Tyler. "We each have to live that story. We each have to validate our own conclusions about life and how the whole thing works, not just as professional scientists, but as individuals who come to active conclusions based on what makes sense and which experiments in life we choose to listen to. And what's great about this is that personal discovery of the scientific method can be as powerful as the traditional myths, at their best, were reputed to be. Like the death and rebirth of Jesus: Joseph Campbell and Jean Houston point out that this is not just something that happened a couple thousand years ago to one very special guy; it is supposed to be a psychological model for the death and rebirth of everyone's ego. I imagine that the celebrations of Marduk slaying Tiamat in the Sumerian rituals were probably great psychological transformers. So let's fantasize: Let's imagine what happens to rituals when the awesome fruits of science start working their way into churches and synagogues and mosques."

"So we're beginning to evolve the existing religions into the larger story told by science," I surmise.

"Yes, and these traditions start inventing new rituals, or revisions of old rituals, to capture the new ideas."

"They do the Tiamat story!" I begin to fantasize.

"I envision a congregation doing a gaian ritual, deep breathing with the planet," he continues. "Now, imagine that these practices start getting locked in. People start having powerful experiences with a particular ritual or celebration. What if the science starts shifting but the people really like these celebrations and want to keep them going?"

"Hmmm. You and I might be able to adjust to a 'just lucky I guess' explanation for the Earth system, but what about people who come from a mind-set where Gaia as the Great Mother has become just too important?"

Tyler responds, "So this means that we really must build into these rituals not just the celebrations of the ideas but a celebration of the fascinating ways we came to those ideas and how we keep moving nearer and nearer the truth. That means that rather than just celebrating the new cosmology, the new geophysiology, there would be instances of celebrating, say, this week's top story in *Nature* and the story of how that new story came to be. Keeping current means we would be celebrating the story of the changing story, too."

"The story of the changing story," I repeat slowly. "That's great! Not only would that approach solve the problem Maynard Smith pointed out; it would positively enhance society's regard for science and for more science."

"One more thing, Julian. Here we are today, a little group of us all excited about catalyzing greater awareness of the evolutionary epic—the evolution of the biosphere as well as all the forms of life. We want to do this in such a way that religious depth and vigor can emerge from a science-based worldview. Yet that's exactly what you tried to do half a century ago—with little, I dare say, to show for it. Why should things be any different today? Will all our energy just be sucked into a black hole of societal indifference?"

"Dark thoughts, dark thoughts," he responds, shaking his head.

"You know I asked your cousin the same question. Here you were, esteemed British biologist, winner of the Royal Society's Darwin Medal, grandson of Thomas Henry Huxley, prize-winning writer of philosophically laced and exceedingly accessible science books, first director-general of the United Nations Educational, Scientific, and Cultural Organization . . ."

"Please!"

"Okay, so I asked your cousin Crispin . . ."

"Ah, Aldous and I used to favor that one, we did. I remember giving him a terrapin for a pet one Christmas in the 1930s. He was so delighted." Huxley pauses in a reverie of memories. Then he returns. "Crispin Tickell became the diplomat of the family; he was well launched on that career path when I died."

"Since then he has moved more and more onto a path converging with your own. He is now head of a graduate college at Oxford University—your alma mater—and thick into the nexus of science, environment, and politics in England. This year, in fact, he delivered the prestigious Mishcon Lecture at University College in London, and the theme he chose was 'God and Gaia: Religion and the Environment.' "

"Well, well!"

"Anyway, to make a long story short, I asked Crispin why it was that your own promotion of the evolutionary epic . . ."

"Marvelous term, *evolutionary epic*! Is that your creation?"

"No, that's Ed Wilson's term. We're all using it now."

"Excellent! You were saying . . ."

"Yes. I asked Crispin why it is that your own brand of 'religion without revelation' is virtually unknown to my generation. He didn't have an answer. Yet I need an answer, because in order for us to have a better chance at success this time around, we must understand why you failed."

"My dear," his voice was dripping with disdain. "I did not fail. My society failed me." He gazes off into the distance. "Poor timing, on my part, to have been born when I was."

"Seriously, Julian, do you think we have a chance of being even a bit successful this time around?"

"A chance? My dear, if you and your friends fail, you will all have to suffer my wrath. The timing couldn't be any more perfect. I am green with envy. You see, it's not a matter of finding a crack in the spiritual ethos and then slipping in the evolutionary epic, as I tried to do. Your society has a bloody gaping hole; its vigor is being drained for want of something to fill it. And the evolutionary epic is just the thing; it may be the only thing."

Catching on, I ask, "Is that gaping hole perchance in need of something green to fill it?"

"Precisely. That gaping hole can be plugged only by a value system through which Western civilization recognizes its debt to and reciprocity with the very environment that sustains it. And value systems, in turn, emerge only from a cosmology. Ergo, the need for our epic."

"So you're with us on the angle we're taking this time around?"

Huxley looked disappointed that I would even have to ask. "Who, my dear, was the first director-general of UNESCO? Who launched the International Union for the Conservation of Nature? Who chaired the committee in 1947 that initiated Britain's exemplary system of Nature Reserves? Who spent a part of his career at the helm of the London Zoo?"

"How could I have forgotten!"

"Yes, of course, I am with you! Give my regards to Loyal and Ursula and Ed and Brian and Mary Evelyn and the rest of the gang. You all must carry on!"

He waved his arm, signaling an end to our conversation, and began to fade. But his voice held on for one more thought.

"Oh yes, and do ring me up again sometime soon, won't you? I love a good conversation—come to think of it, conversation is a nice metaphor for the evolutionary process, don't you think?"

"I'll keep that in mind."

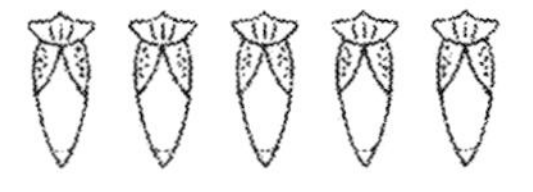

Endnotes

Chapter 1 The Way of Science

1 Quotation by Arne Naess in his 1995 "Self-realization: An ecological approach to being in the world," in George Sessions, ed., *Deep Ecology for the 21st Century* (Boston: Shambala), p. 229. See also Arne Naess, 1989, *Ecology, Community, and Lifestyle*, translated and edited by David Rothenberg (Cambridge: Cambridge University Press), pp. 85–86.

2 Joseph Campbell described religion and religious myths as that which puts one "in accord" with the universe in many of his books. See, for example, Joseph Campbell with Bill Moyers, 1988, *The Power of Myth* (New York: Doubleday), pp. 55–56, 70. Campbell criticizes the misrepresentation of religious metaphor as material fact (pp. 56, 218–19).

3 Huston Smith defines religion as "that which gives meaning to the whole" in his 1991/1958 *The World's Religions* (San Francisco: HarperCollins), p. 388.

4 Lawrence Kohlberg defined religion in his 1981 *The Philosophy of Moral Development* (San Francisco: Harper & Row), p. 321.

5 James Gustafson defines religion in his 1995 "Tracing a trajectory," *Zygon* 30: 177–190.

6 Loyal Rue defines religion as "an integrated understanding of how things are and what things matter" in his opening (unpublished) talk for the 1996 conference "The Epic of Evolution," the 43rd annual meeting of the Institute on Religion in an Age of Science, at Star Island, New Hampshire. See also Loyal Rue, 1989, *Amythia: Crisis in the Natural History of Western Culture* (Tuscaloosa: University of Alabama Press).

7 Quotations by Jacques Monod in his 1971, *Chance and Necessity* (New York: Knopf), pp. 167, 169.

8 Edward O. Wilson, 1975, *Sociobiology: The New Synthesis* (Cambridge, Mass.: Harvard University Press).

9 Quotation by Edward O. Wilson in his 1978 *On Human Nature* (Cambridge, Mass.: Harvard University Press), p. 169. The quotations that immediately follow are drawn from pp. 172, 201.

10 Quotations by Richard Dawkins in his 1976 *The Selfish Gene* (Oxford: Oxford University Press).

11 Quotation by Edward O. Wilson in his 1984 *Biophilia* (Cambridge, Mass.: Harvard University Press), p. 1.

12 The beginnings of a research program on biophilia are evident in Stephen R. Kellert and Edward O. Wilson, 1993, *The Biophilia Hypothesis* (Washington, D.C.: Island Press).

13 Jared Diamond expresses doubt about biophilia in his 1993 "New Guineans and their natural world," in Stephen R. Kellert and Edward O. Wilson, eds., *The Biophilia Hypothesis*, pp. 251–271.

14 The term *exaptation* was coined in Stephen Jay Gould and Elizabeth Vrba, 1981, "Exaptation: A missing term in the science of form," *Paleobiology* 8: 4–15.

15 The exaptation of insect wings from sails is presented in James H. Marden and Melissa G. Kramer, 1995, "Locomotor performance of insects with rudimentary wings," *Nature* 377: 332–34. See also James H. Marden, 1995, "How insects learned to fly," *The Sciences*, November, pp. 26–30.

16 The exaptation of the vertebrate jaw from gill arches is presented in J. Mallatt, 1996, "Ventilation and the origin of jawed vertebrates," *Zoological Journal of the Linnean Society* 117: 329–404. See also Carl Zimmer, 1998, *At the Water's Edge: The Science of Macroevolution Comes of Age* (New York: Simon & Schuster / Free Press).

17 Quotation by Richard Nelson in his 1991/1989 *The Island Within* (New York: Vintage), p. 123.

18 For a discussion of how communing with the One may be followed by compassion for the Many and an ignited activism, see Ken Wilber, 1995, *Sex, Ecology, Spirituality: The Spirit of Evolution* (Boston: Shambala), p. 310.

19 Quotation by William James in his 1902 *Varieties of Religious Experience* (New York: Longmans, Green & Co.), p. 487.

20 Quotation by Gary Snyder in the "Four changes" chapter of his 1974 *Turtle Island* (New York: New Directions).

21 Mircea Eliade offers the idea of *Homo religiosus* in many of his books.

22 I edited two anthologies of essays on the broad theme of meaning in evolutionary biology, both published by MIT Press (Cambridge, Mass.). These are *From Gaia to Selfish Genes: Selected Writings in the Life Sciences* (1991) and *Evolution Extended: Biological Debates on the Meaning of Life* (1994).

Chapter 2 Science and the Coming of a New Story

1 Quotation by Edward O. Wilson on the evolutionary epic in his 1978 *On Human Nature* (Cambridge, Mass.: Harvard University Press), p. 201.

2 Edward O. Wilson, 1992, *The Diversity of Life* (Cambridge, Mass.: Harvard University Press).

3 The exact wording of the last two paragraphs of Wilson's *The Diversity of Life* is as follows:

"The evidence of swift environmental change calls for an ethic uncoupled from other systems of belief. Those committed by religion to believe that life was put on earth in one divine stroke will recognize that we are destroying the Creation, and those who perceive biodiversity to be the product of blind evolution will agree. Across the other philosophical divide, it does not matter whether species have independent rights or, conversely, that moral reasoning is uniquely a human concern. Defenders of both premises seem destined to gravitate toward the same position on conservation.

"The stewardship of environment is a domain on the near side of metaphysics where all reflective persons can surely find common ground. For what, in the final analysis, is morality but the command of conscience seasoned by a rational examination of consequences? And what is a fundamental precept but one that serves all generations? An enduring environmental ethic will aim to preserve not only the health and freedom of our species, but access to the world in which the human spirit was born."

4 Pierre Teilhard de Chardin presents evolution as a "groping" of life toward the Omega Point in his 1959 *The Phenomenon of Man* (New York: Harper & Row).

5 Julian Huxley presented his view of the evolutionary epic in many books. Highlights of his writings are presented in Connie Barlow, ed., 1994, *Evolution Extended: Biological Debates on the Meaning of Life* (Cambridge, Mass.: MIT Press).

6 Daniel Dennett presents the "cranes" versus "skyhooks" view of evolution in his 1995 *Darwin's Dangerous Idea* (New York: Simon & Schuster), pp. 74–76, 144.

7 Mary Catherine Bateson, 1989, *Composing a Life* (New York: Atlantic Monthly Press). See also Joseph Campbell with Bill Moyers, 1988, *The Power of Myth* (New York: Doubleday), p. 229. For a discussion on whether one lives a narrative life or simply recalls it as a narrative in hindsight, see Alasdair MacIntyre, 1981, *After Virtue* (Notre Dame, Indiana: University of Notre Dame), pp. 197–202.

8 Gordon Kaufman depicts the evolutionary process as "serendipitous creativity" in his 1993 *In Face of Mystery: A Constructive Theology* (Cambridge, Mass.: Harvard University Press), p. 279.

9 Loren Eiseley characterized evolution as "the immense journey" in his 1946 *The Immense Journey* (New York: Random House).

10 Quotations by Brian Swimme on Copernicus and the sun appear in his 1996 video *The Hidden Heart of the Cosmos* (Mill Valley, Calif.: Center for the Story of the Universe). A companion volume was published by Orbis (Maryknoll, N.Y.).

11 The star Tiamat is named by Brian Swimme and Thomas Berry in their 1992 *The Universe Story* (San Francisco: Harper San Francisco). See also Swimme's 1984 *The Universe Is a Green Dragon: A Cosmic Creation Story* (Santa Fe: Bear & Co.).

12 Brian Swimme's quotations on Tiamat appear in *The Universe Story*, pp. 8, 60–61.

13 Theodosius Dobzhansky extolled "unity in diversity" in his 1967 *The Biology of Ultimate Concern* (New York: New American Library).

14 The reenactment of speciation in the genus *Helianthus*, by way of hybridization between two pre-existing species, is reported in Loren H. Reiseberg *et al.*, 1996, "Role of gene interactions in hybrid speciation: Evidence from ancient and experimental hybrids," *Science* 272: 741–44. A review article by Jarry Coyne, "Speciation in action," appears in the same issue, pp. 700–701. See also Carol Kaesuk Yoon, 1996, "First ever re-creation of new species birth," *The New York Times*, May 7, pp. C1, C17.

15 Alister Hardy discusses convergent evolution in his 1965 *The Living Stream: Evolution and Man* (New York: Harper & Row), pp. 199–202. For a fascinating look at convergence in placental and marsupial digestive systems, see Ian Hume, 1989, "Reading the entrails of evolution," *New Scientist*, April 15, pp. 43–47.

16 George Gaylord Simpson discusses convergence in vertebrate wings in his 1949 *The Meaning of Evolution* (New Haven, Conn.: Yale University Press), p. 67.

17 Richard Dawkins shows the step-by-step functionality of eye evolution in his 1996 *Climbing Mount Improbable* (New York: Norton), chap. 5.

18 The symbiogenetic origin of plastids on three occasions is reviewed in John Maynard Smith and Eörs Szathmáry, 1995, *The Major Transitions in Evolution* (Oxford: Freeman), pp. 141–42.

19 Five-fold convergence of algae and fungi into lichens is presented in Andrea Gargas, Paula T. DePriest, Martin Grube, and Anders Tehler, 1995, "Multiple origins of lichen symbioses in fungi suggested by SSU rDNA phylogeny," *Science* 268: 1492–94. (See also p. 1437 in the same issue for a review article.)

20 For details on the origin of insect wings, see Michalis Averof and Stephen M. Cohen, 1997, "Evolutionary origins of insect wings from ancestral gills," *Nature* 385: 627–30.

21 Stephen Jay Gould consistently presents a trendless evolutionary history in all his books and essays. His latest is no exception: the 1996 *Full House* (New York: Harmony).

22 Gould's "bush of life" metaphor appears in his *Full House*, p. 18.

23 Kevin Kelly's "thicket of life" metaphor appears in his 1994 *Out of Control: The Rise of Neobiological Civilization* (Addison-Wesley), p. 372.

24 Ursula Goodenough's "credo of continuity" appears in her 1994 "The religious dimensions of the biological narrative," *Zygon* 29: 603–18. See also her "What science can and cannot offer to a religious narrative," *Zygon* 29: 321–30.

25 Julian Huxley's depiction of trends in evolution appear in his 1923 *Essays of a Biologist* (New York: Knopf), p. 30. See this and other extracts from Huxley's writings on evolutionary progress in Connie Barlow, ed., 1994, *Evolution Extended: Biological Debates on the Meaning of Life* (Cambridge, Mass.: MIT Press).

26 John Maynard Smith and Eörs Szathmáry present eight major breakthroughs in information transfer in their 1995 *The Major Transitions in Evolution* (Oxford: Freeman). A popular version of this book is forthcoming.

27 Quotations by Richard Dawkins in his 1995 *River Out of Eden: A Darwinian View of Life* (New York: Basic Books), pp. xi, 46.

28 Christian de Duve writes of the seven stages of complexification in his 1995 *Vital Dust: Life as Cosmic Imperative* (New York: Basic).

29 Edward O. Wilson presents his views of trends in evolution in his 1992 *The Diversity of Life* (Cambridge, Mass.: Harvard University Press). Key extracts are reprinted in Connie Barlow, ed., 1994, *Evolution Extended: Biological Debates on the Meaning of Life* (Cambridge, Mass.: MIT Press).

30 For a scientific, yet epic, telling of the evolution of organisms as environments for other organisms, see Mark McMenamin and Dianna McMenamin, 1994, *Hypersea: Life on Land* (New York: Columbia University Press), chap. 8.

31 The quotation by Brian Swimme on predation appears in Swimme and Berry's *The Universe Story*, p. 104.

32 Loyal Rue presents the concept of amythia in his 1989 *Amythia: Crisis in the Natural History of Western Culture* (Tuscaloosa: University of Alabama Press). The two direct quotations are drawn from pages 46 and 2. See also his 1997 "Confessions of a shallow environmentalist," *Earthlight*, summer 1997.

33 Jared Diamond identifies humans as the third species of chimpanzee in his 1992 *The Third Chimpanzee* (New York: HarperCollins). Rather than renaming us *Pan sapiens*, however, he suggests renaming chimps *Homo troglodytes*, p. 25.

34 Thomas Berry presents many of his ideas in his 1988 *The Dream of the Earth* (San Francisco: Sierra Club Books).

35 The quotation by Thomas Berry appears in *Dream of the Earth*, pp. 17–18.

36 The mission statement of the Institute on Religion in an Age of Science is as follows: IRAS is a non-denominational, independent society with two purposes. (1) To formulate dynamic and positive relationships between the concepts developed by science and the goals and hopes of humanity expressed through religion. (2) To state human values and contemporary knowledge in such universal and valid terms that they may be understood by all peoples, whatever their cultural background and experience, and provide a basis for worldwide cooperation. (IRAS is the publisher of *Zygon: Journal of Religion and Science.*)

37 Books and papers by Loyal Rue, Ursula Goodenough, and Brian Swimme were previously cited. Rue is a professor at Luther College in Decorah, Iowa. Goodenough is a professor at Washington University in St. Louis. Swimme is on the faculty of the California Institute for Integral Studies, in San Francisco. Mary Evelyn Tucker and John Grim coedited the 1994 *Worldviews and Ecology* (Maryknoll, N.Y.: Orbis). Tucker and Grim are on the faculty of Bucknell University in Lewisburg, Pennsylvania.

38 Quotation by Brian Swimme in his 1984 *The Universe Is a Green Dragon* (Santa Fe: Bear & Co.).

39 The melody for *The Tiamat Song* is shown here. (*Tiamat* is pronounced TEE-ah-maht.)

Chapter 3 Biology and the Celebration of Diversity

1 New discoveries of mammal species in the past ten years are reviewed in Virginia Morell, 1996, "New mammals discovered by biology's new explorers," *Science* 273: 1491.

2 Edward O. Wilson writes about indigenous knowledge of ants in his 1992 *The Diversity of Life* (Cambridge, Mass: Harvard University Press), p. 43.

3 Brian Swimme presents the tragedy of extinction as a loss of perception in his 1991 *Canticle to the Cosmos* (Petaluma, Calif.: CSU Video), videotape no. 3 of 12.

4 An assessment of imperiled fynbos plants is included in Stuart L. Pimm *et al.*, 1995, "The future of biodiversity," *Science* 269: 347–50.

5 Imperiled pollinators are the subject of Stephen L. Buchmann and Gary Paul Nabhan, 1996, *The Forgotten Pollinators* (Washington D.C.: Island Press).

6 Peter Raven's estimates of linked extinctions is quoted in Paul Ehrlich and Anne Ehrlich, 1981, *Extinction* (New York: Random House), p. 139.

7 Extinction of passenger pigeon lice is reported by Nigel E. Stork, 1993, "Extinction or 'co-extinction' rates?" *Nature* 366: 307.

8 Quotation by Donald A. Windsor in his 1995 "Endangered interrelationships," *Wild Earth*, Winter 95/96, pp. 78–83.

9 List of North American vertebrate extinctions appears in Richard B. Primack, 1993, *Essentials of Conservation Biology* (Sunderland, Mass: Sinauer), pp. 242–43.

10 The U.N. estimates of species on the brink reported in V. H. Heywood, ed., 1995, *Global Biodiversity Assessment* (published for the United Nations Environment Program by Cambridge University Press).

11 The Overkill Hypothesis is the subject of Paul S. Martin and Richard G. Klein, eds., 1984, *Quaternary Extinctions: A Prehistoric Revolution* (Tucson: University of Arizona Press). See also Paul S. Martin, 1990, "40,000 years of extinctions on the 'planet of doom,' " *Palaeogeography, Palaeoclimatology, Palaeoecology* 82: 187–201.

12 Estimates of oceanic island extinctions caused by Polynesian colonization are presented by Storrs L. Olson, 1989, "Extinction on islands: Man as a catastrophe," in David Western and Mary C. Pearl, eds., *Conservation for the Twenty-First Century* (New York: Oxford University Press), pp. 50–53. See also David W. Steadman, 1995, "Extinction of birds on tropical Pacific islands," in David W. Steadman and Jim I. Mead, eds., *Late Quaternary Environments and Deep History: A Tribute to Paul S. Martin* (Hot Springs, S.D.: Mammoth Site of Hot Springs, South Dakota), pp. 33–49.

13 A review of the Madagascar extinctions appears in Elizabeth Culotta, 1995, "Many suspects to blame in Madagascar extinctions," *Science* 268: 1568–69.

14 Jared Diamond's assessment of moral culpability for extinctions appears in his 1992 *The Third Chimpanzee: The Evolution and Future of the Human Animal* (New York: HarperCollins), pp. 336–37.

15 Jared Diamond has a chapter on "The Golden Age that never was" in his 1992 *The Third Chimpanzee.*

16 The quotation by Edward O. Wilson on beginning "the age of restoration" appears in his 1992 *The Diversity of Life* (Cambridge, Mass: Harvard University Press), p. 351.

17 For a presentation of the history of island biogeography theory and its current uses, see David Quammen, 1996, *The Song of the Dodo: Island Biogeography in an Age of Extinctions* (New York: Scribner).

18 Sky islands of the American southwest are described in Tony Povilitis, 1996, "The Gila River sky island bioregion," *Natural Areas Journal* 16: 62–66. See also William K. Stevens, 1996, "Ranchers ride to the defense of mountaintop 'sky islands,'" *The New York Times*, 24 July, pp. C1, C8.

19 Habitat fragmentation of national parks was identified by William Newmark, 1987, "A land-bridge island perspective on mammalian extinctions in western North American parks," *Nature* 325: 430–32. Updated in his 1995 "Extinction of mammal populations in western North American national parks," *Conservation Biology* 9: 512–26.

20 For a summary of monarch butterfly natural history and conservation problems, see Carl Zimmer, 1996, "The flight of the butterfly," *Discover*, May, pp. 38–39.

21 For a description of the Santa Cruz monarch festival, see "The winter palace of the monarchs," in Diane Ackerman, 1995, *The Rarest of the Rare* (New York: Random House).

22 The quotations by Diane Ackerman appear in her 1995 *The Rarest of the Rare* (New York: Random House), pp. 131, 27.

23 Diane Ackerman's poem, "Ode to the Alien," appears in her 1991 *Jaguar of Sweet Laughter: New and Selected Poems* (New York: Random House), pp. 189–91.

24 Edward O. Wilson, 1994, *Naturalist* (Washington, D.C.: Island Press).

25 The guidebook for understanding and staging a workshop is *Thinking Like a Mountain: Towards a Council of All Beings*, by John Seed, Joanna Macy, Pat Fleming, and Arne Naess, 1988 (Philadelphia: New Society Publishers).

26 The quotation by John Seed is reprinted in 1985, "Anthropocentrism," Appendix E in Bill Devall and George Sessions, eds., *Deep Ecology* (Layton, Utah: Gibbs Smith Publishing).

Chapter 4 Ecology and the Birth of Bioregionalism

1 The quotation by Gary Snyder appears in his 1990 *The Practice of the Wild* (San Francisco: North Point Press), p. 109.

2 Paul Shepard presented ecology as the science in which "relationships are as real as the thing" in his coauthored book with Daniel McKinley, 1969, *The Subversive Science* (Boston: Houghton Mifflin), p. 3.

3 Gregory Bateson writes of "organism-in-environment" in his 1972 *Steps to an Ecology of Mind* (New York: Ballantine).

4 Key early publications promoting the bioregional movement are Peter Berg, ed., 1978, *Reinhabiting a Separate Country: A Bioregional Anthology of Northern California* (San Francisco: Planet Drum Foundation); and Stephanie Mills, 1981, "Planetary Passion," *Coevolution Quarterly* 32: 4.

5 The quotation by David Abram is in his 1996 *The Spell of the Sensuous* (New York: Pantheon), p. 271.

6 The quotation by J. Baird Callicott is in his 1987 *A Companion to* A Sand County Almanac (Madison: University of Wisconsin Press), p. 163.

7 The rivers of Alabama are presented as a hotspot for mussels in Charles Lydeard and Richard L. Mayden, 1995, "A diverse and endangered aquatic ecosystem of the southeast United States," *Conservation Biology* 9: 800–805.

8 A list of endangered ecosystems appears in R. F. Noss, E. T. LaRoe III, and J. M. Scott, 1995, *Endangered Ecosystems of the United States: A Preliminary Assessment of Loss and Degradation*, Biological Report 28, U.S. Department of Interior, National Biological Service (Washington, D.C.: USGPO). See also Reed Noss, 1995, "What should endangered ecosystems mean to the Wildlands Project?" *Wild Earth*, Winter 95/96.

9 An excellent review article of the keystone species concept is Mary E. Power *et al.*, 1996, "Challenges in the quest for keystones," *BioScience* 46: 609–20.

10 The classic study that identified the starfish as a keystone species is Robert T. Paine, 1966, "Food web complexity and species diversity," *American Naturalist* 100: 65–75.

11 Alien plant species in North America are reported in Daniel Simberloff, 1996, "Impacts of introduced species in the United States," *Consequences* 2: 13–22.

12 The quotation by Paul S. Martin on ghost species appears in his 1992 "The last entire Earth," *Wild Earth*, Winter 92/93, pp. 29–32.

13 Ghost predation of the American pronghorn is reported in Carol Kaesuk Yoon, 1996, "Pronghorn's speed may be legacy of past predators," *The New York Times*, December 24, pp. C1, C6.

14 Paul Martin's publications on extinct Pleistocene mammals and the Overkill Hypothesis include Paul S. Martin and Richard G. Klein, eds., 1984, *Quaternary Extinctions: A Prehistoric Revolution* (Tucson: University of Arizona Press); and Martin's 1990 "40,000 years of extinctions on the 'planet of doom.' " *Palaeogeography, Palaeoclimatology, Palaeoecology* 82: 187–201.

15 The quotation by Paul Martin on the Grand Canyon appears in his 1992 essay in *Wild Earth*.

16 The speculation that condors held on in California because of beached whales appears in Paul Martin's 1990 "40,000 years of extinctions."

17 An excellent essay on why the loss of a keystone species (such as a top predator) requires humans to intervene in nature reserves is Jared Diamond, 1992, "Must we shoot deer to save them?" *Natural History*, August, pp. 2–8.

18 Discussion of the shift to an imbalance-of-nature view is in Stewart T. A. Pickett, V. Thomas Parker, and Peggy L. Fiedler, 1992, "The new paradigm in ecology," in Peggy L. Fiedler and Subodh K. Jain, eds., *Conservation Biology* (New York: Chapman and Hall), pp. 65–83.

19 A much-cited book that is leading the shift toward an imbalance-of-nature paradigm is Daniel Botkin's 1990 *Discordant Harmonies: A New Ecology for the Twenty-First Century* (New York: Oxford University Press). The quotation is drawn from p. 62.

20 The quotation by George Wuerthner on "ecological time warp" is in his 1989 "Musing on island biogeography," *Earth First!* November 1.

21 The quotation by Dave Foreman on the importance of wilderness appears in his 1991 *Confessions of an Ecowarrior* (New York: Crown), pp. 3, 7.

22 The Wildlands Project was initially presented in a special "Wildlands Project" issue of *Wild Earth* magazine in 1992. An update appears in the winter 95/96 issue of the same magazine. A review is in Charles C. Mann and Mark L. Plummer, 1993, "The high cost of biodiversity," *Science* 260: 1868–71.

23 The quotation by Dave Foreman on wilderness and generosity of spirit appears in his 1994 "Where man is a visitor," in David Clarke Burks, ed., *Place of the Wild: A Wildlands Anthology* (Washington, D.C.: Island Press), pp. 225–35.

24 Terry Tempest Williams writes of her visit to Pelham Bay Park in her 1994 "Water Songs," in David Clarke Burks, ed., *Place of the Wild* (Washington D.C.: Island Press), pp. 28–33.

Chapter 5 Geophysiology and the Revival of Gaia

1 The four poetic quotations describing Earth from space are, in order, Carl Sagan, 1973, *Cosmic Connection* (New York: Doubleday), p. 60; Gary Snyder, 1984, *Good, Wild, Sacred* (Madley, Hereford: Five Seasons Press), last page of unpaginated; Diane Ackerman, 1991, *A Natural History of the Senses* (New York: Random House), p. 183; and Edgar Mitchell (astronaut), quoted in Kevin W. Kelley, ed., 1988, *The Home Planet* (Reading, Mass.: Addison-Wesley).

2 Quotation by James Lovelock in his 1979 *Gaia: A New Look at Life on Earth* (Oxford: Oxford University Press), p. 12.

3 The quotation by James Lovelock on the atmosphere appears in his 1979 *Gaia*, p. 10.

4 Evidence of life in martian meteorites reported in David S. McKay *et al.*, 1996, "Search for past life on Mars: Possible relic biogenic activity in martian meteorite AL84001," *Science* 273: 924–30. See also the summary article, pp. 864–66, of same.

5 The existence of a biosphere within Earth was proposed by Thomas Gold in his 1992 "The deep, hot biosphere," *Proceedings of the National Academy of Sciences* 89: 6045–49. See also his forthcoming book by the same title, to be published by Copernicus in 1998.

6 Quotation by James Lovelock in his 1991 "Geophysiology: The Science of Gaia," in Stephen H. Schneider and Penelope J. Boston, eds., *Scientists on Gaia* (Cambridge, Mass.: MIT Press), pp. 3–10.

7 The paradox of a less luminous early sun was proposed by Carl Sagan and George Mullen, 1972, "Earth and Mars: Evolution of atmospheres and surface temperatures," *Science* 177: 52–56. The paradox was named Faint Young Sun by Michael J. Newman and Robert T. Rood, 1977, "Implications of solar evolution for the Earth's early atmosphere," *Science* 198: 1035–37. See also James F. Kasting, 1993, "New spin on ancient climate," *Nature* 364: 759.

8 High volumes of carbon dioxide in Earth's early atmosphere are supported by James F. Kasting in his 1993 "Earth's early atmosphere," *Science* 259: 920–25. For a contrary view that proposes far lower concentrations of carbon dioxide, see Rob Rye, Phillip H. Kuo, and Heinrich D. Holland, 1995, "Atmospheric carbon dioxide concentrations before 2.2 billion years ago," *Nature* 378: 603–605.

9 David Schwartzman and Tyler Volk proposed a biotic role in temperature history in their 1989 "Biotic enhancement of weathering and the habitability of Earth," *Nature* 340: 457–60. See also David Schwartzman and Tyler Volk, 1991, "Biotic enhancement of weathering and the surface temperatures on Earth since the origin of life," *Palaeogeography, Palaeoclimatology, Palaeoecology* 90: 357–71.

10 The end of the biosphere is estimated by James Lovelock and Michael Whitfield in their 1982 "Life span of the biosphere," *Nature* 296: 561–63. A newer estimate is given in Ken Caldeira and James F. Kasting, 1992, "The life span of the biosphere revisited," *Nature* 360: 721–23. See also the review article by Tyler Volk, "When climate and life finally devolve," in the same issue, p. 707.

11 The idea that plankton might influence cloud formation was proposed in R. Charlson, J. E. Lovelock, M. O. Andreae, and S. G. Warren, 1987, "Oceanic phytoplankton, atmospheric sulphur, cloud albedo, and climate, "*Nature* 326: 655–61.

12 The idea that bogs might trigger ice ages is presented by Lee F. Klinger, 1991, "Peatland formation and ice ages: A possible gaian mechanism re-

lated to community succession," in S. H. Schneider and P. J. Boston, eds., *Scientists on Gaia* (Cambridge, Mass.: MIT Press), pp. 247–55.

13 The quotation by James Lovelock on Daisyworld appears in his 1988 *Ages of Gaia* (New York: Norton), pp. 34–35.

14 The first results of Daisyworld modeling were presented in Andrew J. Watson and James E. Lovelock, 1983, "Biological homeostasis of the global environment: The parable of Daisyworld," *Tellus* 35B: 284–89.

15 Scientists who find Daisyworld modeling important include Peter Saunders, 1996, "Daisyworld and the Future of Gaia," in Peter Bunyard, ed., *Gaia in Action* (Edinburgh: Floris Books), pp. 75–88. See also Stephan Harding in Stephan P. Harding and James E. Lovelock, 1996, "Exploiter-mediated coexistence and frequency-dependent selection in a numerical model of biodiversity," *Journal of Theoretical Biology* 182: 109–16.

16 A critic of Gaia and Daisyworld is George C. Williams, in his 1992 "Gaia, nature worship, and biocentric fallacies," *Quarterly Review of Biology* 67: 479–86.

17 Quotation by Stanley Salthe in his 1993 "What implications does Gaia pose for evolutionary biology?" Prepared remarks for a panel discussion, "Problems of Gaia in Evolutionary Biology," organized by Connie Barlow for the biennial meeting of the International Society for the History, Philosophy, and Social Studies of Biology," Evanston, Illinois. See also Salthe's 1993 book, *Development and Evolution: Complexity and Change in Biology* (Cambridge, Mass.: MIT Press).

18 Stuart Kauffman's views on a self-organizing universe are presented in his 1995 *At Home in the Universe: The Search for Laws of Self-Organization and Complexity* (New York: Oxford University Press).

19 The quotation by Michael Rampino appears in his 1991 "Gaia versus Shiva: Cosmic effects on the long-term evolution of the terrestrial biosphere," in *Scientists on Gaia*, pp. 382–90. See also Michael Rampino and Bruce M. Haggerty, 1996, "The Shiva hypothesis: Impacts, mass extinctions, and the galaxy," *Earth, Moon, and Planets* 72: 441–60.

20 A causal link between the end-Cretaceous Yucatán crater and the Deccan flood basalt is presented in Michael Rampino and Ken Caldeira, 1992, "Antipodal hotspot pairs on the Earth," *Geophysical Research Letters* 19: 2011–14.

21 The phrase "cataclysms galore" appears in the poem "Mars" by Diane Ackerman in her 1991 *The Jaguar of Sweet Laughter: New and Selected Poems* (New York: Random House).

22 For discussion of the ecological changes wrought by the evolution of fecal pellets, see Malcolm Walter, 1995, "Faecal pellets in world events," *Nature* 376: 16. See also Lori Oliwenstein, 1996, "Life's grand explosions," *Discover*, January, pp. 42–43.

23 A possible atmospheric imbalance caused by the evolution of lignin is proposed by Jennifer M. Robinson, 1991, "Phanerozoic atmospheric reconstructions," *Palaeogeography, Palaeoclimatology, Palaeoecology (Global and Planetary Change Section)* 97: 51–62.

24 Connie Barlow and Tyler Volk, 1990, "Open systems living in a closed biosphere: A new paradox for the Gaia debate," *Biosystems* 23: 371–84.

25 The recipe for building a baby biosphere is in Stephen Tomkins, 1995, "Science for the Earth starts at school," in Tom Wakeford and Martin Walters, eds., *Science for the Earth* (New York: Wiley), pp. 257–76.

26 The original version of this updated quotation by Lynn Margulis appears in her 1995 "A pox called man," in *Science for the Earth*, pp. 19–36.

27 The early Gaia papers are Lynn Margulis and James E. Lovelock, 1974, "Biological modulation of the earth's atmosphere," *Icarus* 21: 471–89; James E. Lovelock and Lynn Margulis, 1974, "Atmospheric homeostasis by and for the biosphere: The Gaia hypothesis," *Tellus* 26: 2–9; and James E. Lovelock and Lynn Margulis, 1974, "Homeostatic tendencies of the Earth's atmosphere," *Origins of Life* 5: 93–103.

28 Hinkle's remark, "Gaia is symbiosis as seen from space," appears in Lynn Margulis, 1993/1981, *Symbiosis in Cell Evolution: Microbial Communities in the Archean and Proterozoic Eons*, 2nd ed. (New York: Freeman), p. 367.

29 For more on the microbial perspective, see Lynn Margulis, 1995, "Gaia is a tough bitch," in John Brockman, *The Third Culture*, (New York: Simon & Schuster), pp. 129–51.

30 The "population of one" criticism of Gaia has been raised by Richard Dawkins, 1982, *The Extended Phenotype* (New York: Freeman), pp. 234–37. See also W. F. Doolittle, 1981, "Is nature really motherly?" *Coevolution Quarterly* 29: 58–65.

31 The quoted biographer of Vernadksy is Rafal Serafin, 1988, "Noosphere, Gaia, and the science of the biosphere," *Environmental Ethics* 10: 121–37.

32 Until very recently, the only English-language books on Vernadksy's ideas were translations from two Russian books: M. M. Kamshilov, 1976, *Evolution of the Biosphere* (Moscow: Mir); and Andrei Lapo, 1982, *Traces of Bygone Biospheres* (Moscow: Mir). In 1997, Vernadsky's *The Biosphere* was finally made available in English, as *The Biosphere: Complete Annotated Edition*, translated by David Langmuir, edited and annotated by Mark McMenamin, with an introduction by Jacques Grinevald and a foreword by Lynn Margulis *et al.* (New York: Copernicus).

33 The quotation by Margulis on albatrosses appears in her chapter in *Gaia in Action*, p. 63.

34 Margulis contrasts Lovelock's and Vernadsky's perspectives in Lynn Margulis and Dorion Sagan, 1995, *What Is Life?* (New York: Simon & Schuster), p. 44.

35 For Vernadsky on the realm of "bygone biospheres," see Andrei Lapo, *Traces of Bygone Biospheres*.

36 For Schwartzman's view of the temperature history of Earth, see David Schwartzman and Tyler Volk, 1991, "Biotic enhancement of weathering and the surface temperatures of Earth since the origin of life," *Palaeogeography, Palaeoclimatology, Palaeoecology* 90: 357–71.

37 Michael Rampino's paradigm-shifting idea appears in his 1994 "Tillites, diamictites, and ballistic ejecta of large impacts," *Journal of Geology* 102: 439–56. See also V. R. Oberbeck, J. R. Marshall, and H. Aggarwal, 1993, "Impacts, tillites, and the breakup of Gondwanaland," *Journal of Geology* 101: 1–19.

38 The quotation by Maynard Smith appears in John Maynard Smith and Eörs Szathmáry, 1995, *The Major Transitions in Evolution* (Oxford: Freeman), p. 145.

39 David Schwartzman and George C. Williams participated in the discussion I organized, "Problems of Gaia in evolutionary biology," at the 1993 biennial meeting of the International Society for the History, Philosophy, and Social Studies of Biology, in Evanston, Illinois.

40 The idea that evolutionary events happened as soon as temperatures allowed appears in David Schwartzman, Mark McMenamin, and Tyler Volk, 1993, "Did surface temperatures constrain microbial evolution?" *BioScience* 43: 390–93.

41 The spiral image of biospheric evolution appears in M. M. Kamshilov, *Evolution of the Biosphere* (Moscow: Mir), p. 203.

42 Gary Snyder stated, "We are all indigenous to this planet," in oral remarks at the "Watershed Conference" of the Orion Society and the Library of Congress, 1996, Washington, D.C.

43 The marine N and P puzzle was identified by Alfred C. Redfield in his 1958, "The biological control of chemical factors in the environment," *American Scientist* 46: 205–21.

44 The Vernadksy paper mentioned by Schwartzman is 1945, "The biosphere and the noosphere," *American Scientist* 33: 1–12. Vernadsky's *Biosphere* has just been translated into English (see note 32).

45 David Schwartzman's key papers are cited in notes 9, 40, and 53. See also David Schwartzman and S. N. Shore, 1996, "Biotically mediated surface cooling and habitability for complex life," in L. R. Doyle, ed., *Circumstellar Habitable Zones* (Menlo Park, Calif.: Travis House), pp. 421–43.

46 The Daisyworld article mentioned by Volk is James E. Lovelock, 1983, "Daisyworld: A cybernetic proof of the Gaia hypothesis," *CoEvolution Quarterly*, Summer, pp. 66–72.

47 Tyler Volk's papers mentioned in the text include his 1987 "Feedbacks between weathering and atmospheric CO_2 over the last 100 million years," *American Journal of Science* 287: 763–79; Volk, 1989, "Rise of angiosperms

as a factor in long-term climatic cooling," *Geology* 17: 107–10; Tyler Volk, 1989, "Sensitivity of climate and atmospheric CO_2 to deep-ocean and shallow-ocean carbonate burial," *Nature* 337: 637–40; Connie Barlow and Tyler Volk, 1990, "Open systems living in a closed biosphere: A new paradox for the Gaia debate," *BioSystems* 23: 371–84; and Barlow and Volk, 1992, "Gaia and evolutionary biology," *BioScience* 42: 686–93.

48 Klinger is named "Bog Man" in Carl Zimmer, 1991, "Bog Man," *Discover*, April, pp. 62–67.

49 Lee Klinger's papers include his 1991 "Peatland formation and ice ages: A possible gaian mechanism related to community succession," in *Scientists on Gaia*, pp. 247–55; Lee F. Klinger, John A. Taylor, and Lars G. Franzen, 1996, "The potential role of peatland dynamics in ice-age initiation," *Quaternary Research* 45: 89–92; Klinger, 1996, "The myth of the classic hydrosere model of bog succession," *Arctic and Alpine Research* 28: 1–9; and Klinger, 1990, "Global patterns in community succession: Bryophytes and forest decline," *Memoirs of the Torrey Botanical Club* 24(1): 1–50.

50 The article that triggered Watson's interest is James E. Lovelock and Sidney Epton, 1974, "Quest for Gaia," *New Scientist*, February, pp. 304–306.

51 Andrew Watson's papers include A. Watson, J. E. Lovelock, and L. Margulis, 1978, "Methanogenesis, fires, and the regulation of atmospheric oxygen," *BioSystems* 10: 293–98; A. Watson and J. E. Lovelock, 1983, "Biological homeostasis of the global environment: The parable of Daisyworld," *Tellus* 35B: 284–89; and A. Watson *et al.*, 1994, "Minimal effect of iron fertilization on sea-surface carbon dioxide concentrations," *Nature* 371: 143–45.

52 Lee Kump's Gaia papers include Lee R. Kump and James E. Lovelock, 1995, "The geophysiology of climate," in A. Henderson-Sellers, ed., *Future Climates of the World: A Modeling Perspective* (Oxford: Elsevier), pp. 537–53; and James E. Lovelock and Lee R. Kump, 1994, "Failure of climate regulation in a geophysiological model," *Nature* 369: 732–34.

53 Schwartzman's paper on SETI and Gaia is David Schwartzman and L. J. Rickard, 1988, "Being optimistic about SETI," *American Scientist* 76: 364–69.

54 Harding's paper with Lovelock is Stephan P. Harding and James E. Lovelock, 1996, "Exploiter-mediated coexistence and frequency-dependent selection in a numerical model of biodiversity," *Journal of Theoretical Biology* 182: 109–16.

Chapter 6 Meaning-Making

1 Ursula Goodenough discusses the emergence of meaning with the very first cell in her 1994 "The religious dimensions of the biological narrative," *Zygon* 29: 603–18.

2 Quotation by Holmes Rolston III in his 1994 *Conserving Natural Value* (New York: Columbia University Press), p. 168.

3 Quotation by James E. Lovelock in his 1979 *Gaia: A New Look at Life on Earth* (Oxford: Oxford University Press), p. 10.

4 For a discussion of the "breathing of the biosphere," see Tyler Volk, 1997, *Gaia's Body: Toward a Physiology of Earth* (New York: Copernicus), chap. 1.

5 Quotations by Bryan Appleyard in his 1992 *Understanding the Present: Science and the Soul of Modern Man* (New York: Doubleday), pp. 76, 101, 107.

6 The poem by Joy Harjo is an extract drawn from p. 56 of *Secrets from the Center of the World*, by Joy Harjo and Stephen Strom (volume 17 of Sun Tracks: An American Indian Literary Series), copyright 1989 by the Arizona Board of Regents. The version that appears here was modified slightly by the poet and reprinted (with line ends determined by the editor and approved by the poet) in my own 1994 *Evolution Extended: Biological Debates on the Meaning of Life* (Cambridge, Mass.: MIT Press), p. 293.

7 For more on meaning-making, see Rodney Holmes, 1996, "*Homo religiosus* and its brain: Reality, imagination, and the future of nature," *Zygon* 31: 441–56. See also James B. Ashbrook, 1996, "Interfacing religion and the neurosciences: A review of twenty-five years of exploration and reflection," *Zygon* 31: 545–82.

8 Loyal Rue spoke of religion as "how things are and what things matter" in his (unpublished) introductory talk at the 1996 "Epic of Evolution" conference of the Institute on Religion in an Age of Science, Star Island, N.H.

9 Tyler Volk discusses the metaphysical binaries from a science perspective in his 1995 *Metapatterns Across Space, Time, and Mind* (New York: Columbia University Press), chap. 4.

10 Philemon Sturges, in a personal communication, spoke of the end Cretaceous as the time when Earth was "struck by a shooting star."

11 Richard Dawkins writes of an arms race among trees in his 1986 *The Blind Watchmaker* (New York: Norton), p. 184.

12 For details on the synergies that come by way of symbiosis, cooperation, and sociality, see Peter Corning, 1983, *The Synergism Hypothesis* (New York: McGraw-Hill).

13 Quotation by Stephen Jay Gould in his 1980 "Is a new and general theory of evolution emerging?" *Paleobiology* 6(1): 119–30.

14 Loyal Rue writes of "a federation of meaning" in his 1994 *By the Grace of Guile* (New York: Oxford University Press), p. 284.

15 Clifford Matthews explains his "mandala for science" in his 1995 "Images of enlightenment: A mandala for science," in Clifford N. Matthews and

Roy Abraham Varghese, eds., *Cosmic Beginnings and Human Ends: Where Science and Religion Meet* (Chicago: Open Court), pp. 11–30.

16 Warwick Fox offers an excellent discussion of "deep questioning" as set forth by the originator of the "deep ecology" ecophilosophy, Arne Naess. See Fox's 1995 *Toward a Transpersonal Ecology* (Albany: SUNY Press), pp. 91–103.

17 Quotation by John Haught in his 1990 *What Is Religion?* (Mahwah, New Jersey: Paulist Press), p. 230.

18 Stephen Jay Gould's latest statement on the boundary between science and religion is his 1997 "Nonoverlapping magisteria," *Natural History*, March, pp. 16–22, 60–62.

19 Peter Kropotkin ably, if too romantically, set forth the scope of cooperation in nature in his 1902 book, *Mutual Aid.* The modern scientific vision of cooperation is embedded in the notion of symbiosis, set forth by Lynn Margulis in her 1993/1981 *Symbiosis in Cell Evolution* (New York: Freeman). See also Peter Corning's 1983 *The Synergism Hypothesis* (New York: McGraw-Hill).

20 For a summary of anthropocentric values in nature, see Stephen R. Kellert, 1996, *The Value of Life* (Washington, D.C.: Island Press).

21 Anthony Weston advocates pluralistic ethics in his 1985 "Beyond intrinsic value: Pragmatism in environmental ethics," *Environmental Ethics* 7: 321–39.

22 Quotation by Dave Foreman in his 1991 *Confessions of an Eco-Warrior* (New York: Crown), p. 9.

23 Quotation by Gary Snyder in his 1974/1969 "Four changes" chapter, *Turtle Island* (New York: New Directions).

24 The duration of biotic recovery following a mass extinction has been estimated by E. O. Wilson in his 1992 "Biophilia and the conservation ethic," in Stephen R. Kellert and E. O. Wilson, eds., *The Biophilia Hypothesis* (Washington, D.C.: Island Press), pp. 31–41. See also V. H. Heywood, 1995, *Global Biodiversity Assessment* (Cambridge: Cambridge University Press), p. 197.

25 Quotation by Gary Snyder in his 1990 *The Practice of the Wild* (Berkeley: North Point Press), p. 176.

26 Holmes Rolston III presents the idea of "storied achievement" in his 1994 *Conserving Natural Value* (New York: Columbia University Press), pp. 174–81.

27 Aldo Leopold's "land ethic" is presented in his 1949 *A Sand County Almanac* (New York: Oxford University Press), pp. 224–25. Baird Callicott supports the land ethic in his 1989 *In Defense of the Land Ethic* (Albany: SUNY Press).

28 Quotation by James Lovelock in his 1991 *Healing Gaia* (New York: Harmony), p. 17.

29 "Because it is my religion" is the title of an essay I wrote for the fall 1996 issue of *Wild Earth*, pp. 5–11. See also my "Re-storying biodiversity" in the spring 1997 issue, pp. 14–18.

30 Holmes Rolston develops the distinction between "spontaneous" nature and "deliberated" culture in his 1994 *Conserving Natural Value* (New York: Columbia University Press), p. 4.

31 The quotation and concepts by J. Baird Callicott on rights extended to nature appear in his 1989 *In Defense of the Land Ethic*, pp. 33, 47, 134–36.

32 Warwick Fox provides an excellent survey of those, like himself, who advocate cultivation of ecological consciousness rather than enaction of new forms of rights; he surveys the writings of Arne Naess and others within the deep ecology movement on the importance of working toward a wider identification with the natural world. See his 1995 *Toward a Transpersonal Ecology* (Albany: SUNY Press), pp. 217–31.

33 Carol Gilligan differentiates a code of justice from a code of caring in her 1982 *In a Different Voice: Psychological Theory and Women's Development* (Cambridge, Mass.: Harvard University Press). For a discussion of the differences between the ecofeminist relational perspective and the deep-ecology identification view, see Jim Cheney, 1987, "Ecofeminism and deep ecology," *Environmental Ethics* 9: 115–45. See also Michael E. Zimmerman, 1987, "Feminism, deep ecology, and environmental ethics," *Environmental Ethics* 9: 21–44.

34 Quentin D. Wheeler argues for phylogenetic distinctiveness to guide conservation priorities in his 1995 "Systematics and biodiversity," *BioScience*, Supplement pp. S21–27. See also R. I. Vane-Wright, C. J. Humphries, and P. H. Williams, 1991, "What to protect? Systematics and the agony of choice," *Biological Conservation* 55: 235–54; and T. L. Erwin, 1991, "An evolutionary basis for conservation strategies," *Science* 253: 750–52.

35 Tijs Goldschmidt recounts the sad tale of the vanishing fish of Lake Victoria in his 1996, *Darwin's Dreampond* (Cambridge, Mass.: MIT Press).

36 Discovery of the Romanian cave is reported in Serban M. Sarbu, Thomas C. Kane, and Brian K. Kinkle, 1996, "A chemoautotrophically based cave ecosystem," *Science* 272: 1953–55. See also Malcolm W. Browne, 1995, "Evolving in the dark, over 5.5 million years," *The New York Times*, September 26, pp. C1, C9.

37 Quotation by Christian de Duve in his 1995 *Vital Dust: Life as Cosmic Imperative* (New York: Basic), p. xvii.

38 Dave Foreman's characterization of wilderness as "the arena of evolution" appears in his 1991 *Confessions of an Eco-Warrior* (New York: Crown), pp. 7, 27.

39 Norman Myers originated the concept of biodiversity "hot spots" in his 1988 "Threatened biotas: 'Hot spots' in tropical forests," *The Environmen-*

talist 8(3): 187–208. Myers expanded this idea in a later issue of same journal, 19(4): 243–56. The eighteen hot spots are reviewed by E. O. Wilson in his 1992 *The Diversity of Life* (Cambridge, Mass.: Harvard University Press), pp. 261–70. Specific hot spots in the United States have been proposed by A. P. Dobson *et al.*, 1997, "Geographic distribution of endangered species in the United States," *Nature* 275: 550–53.

40 The sprouting of an ancient arctic lupine seed buried in permafrost is mentioned in Fred Bruemmer, 1987, "Life upon the permafrost," *Natural History*, April, pp. 31–38.

41 John Rawls presents the "veil of ignorance" criterion in his 1971 *Theory of Justice* (Cambridge, Mass.: Harvard University Press).

42 James Lovelock writes of "meadows of the sea" in his 1991 *Healing Gaia: Practical Medicine for the Planet* (New York: Harmony), p. 110.

43 Quotation by Lynn Margulis in her 1993/1981 *Symbiosis in Cell Evolution* (New York: Freeman), p. 367.

44 The calculation of photosynthetic appropriation was made by Peter M. Vitousek, Paul R. Ehrlich, Anne H. Ehrlich, and Pamela A. Matson, 1986, "Human appropriation of the products of biosynthesis," *BioScience* 36: 368–73.

45 Dave Foreman writes of "humanpox" in his 1991 *Confessions of an Eco-Warrior* (New York: Crown), p. 57.

46 Quotation by Neil Everndon in his 1993/1985 *The Natural Alien* (Toronto: University of Toronto Press), p. 110.

47 Quotation by Bryan Appleyard in his 1992 *Understanding the Present: Science and the Soul of Modern Man* (New York: Doubleday), p. 129.

48 For a discussion of the three types of self-image, see Deane Curtin, 1994, "Dōgen, deep ecology, and the ecological self," *Environmental Ethics* 16: 195–213. See also Jim Cheney, 1987, "Eco-feminism and deep ecology," *Environmental Ethics* 9: 115–45.

49 Quotaton by Martin Buber in his 1970/1937 *I and Thou*, translated by Walter Kaufmann (New York: Scribner's), p. 69.

50 Holmes Rolston III offers "the entwined self" in his 1994 *Conserving Natural Value* (New York: Columbia University Press), p. 156.

51 Reference to Deane Curtin (see note 48).

52 Joanna Macy writes of "eco-self" in her 1990 "The greening of the self," in A. Badiner, ed., *Dharma Gaia* (Berkeley: Parallax Press), p. 56.

53 The quotation by Holmes Rolston III on the extended human vascular system appears in his 1986 *Philosophy Gone Wild* (Buffalo: Prometheus), p. 23. The following quotation on life as a propagating wave, p. 131.

54 Arne Naess develops the concept of a greater Self in his 1995 "Self-realization: An ecological approach to being in the world," in George Sessions, ed., *Deep Ecology for the 21st Century* (Boston: Shambala), pp. 225–39.

55 Quotation by Warwick Fox on identity in his 1995 *Toward a Transpersonal Ecology* (Albany: SUNY), p. 231.

56 Aldous Huxley surveys mystical experience of an expanded self in his 1944 *The Perennial Philosophy* (New York: Harper & Row).

57 An embedded (and embodied) sense of self is espoused by Charlene Spretnak in her conversation with Derrick Jensen, in Jensen's 1995 *Listening to the Land* (San Francisco: Sierra Club Books), p. 49–50.

58 Arthur Koestler's notion of holon is presented in his 1978 *Janus: A Summing Up* (New York: Random House). Tyler Volk applies the concept in his 1997 *Gaia's Body: Toward a Physiology of Earth* (New York: Copernicus), chap. 2. Ken Wilber makes the holon a cornerpiece of his worldview in his 1995 *Sex, Ecology, and Spirituality: The Spirit of Evolution* (Boston: Shambala), pp. 41–65.

59 Ian Barbour traces root metaphors in his 1990 *Religion in an Age of Science* (New York: HarperCollins), p. 219.

60 Colin Tudge makes the point that advocates of reduced population are the least misanthropic of all in his 1996 *The Time Before History* (New York: Scribner), p. 319.

61 The quotation by Holmes Rolston III on "the love of life become conscious of itself" appears in his 1993 "Biophilia, selfish genes, shared values," in Stephen R. Kellert and Edward O. Wilson, eds., *The Biophilia Hypothesis* (Washington, D.C.: Island Press), pp. 381–414. His "an evolutionary ecosystem become conscious of itself" appears in his *Philosophy Gone Wild*, p. 141.

62 The quotation by Julian Huxley on "evolution become conscious of itself" appears in his 1957 *Religion Without Revelation* (New York: Harper & Row), p. 209. Excerpts are reprinted in my own 1994 *Evolution Extended*, p. 233.

63 Quotation by James Lovelock on "Gaia's range of perception" appears in his 1979 *Gaia: A New Look at Life* (Oxford: Oxford University Press), p. 148.

64 The quotation by Joseph Campbell on "the eyes of the Earth" appears in Joseph Campbell with Bill Moyers, 1988, *The Power of Myth* (New York: Doubleday), p. 32. See also his 1986 *The Inner Reaches of Outer Space* (New York: Harper & Row), p. 28.

65 Quotations on human role, as follows: Arne Naess, 1995, "Metaphysics of the treeline," in George Sessions, ed., *Deep Ecology for the 21st Century* (Boston: Shambala), p. 248; Gary Snyder, 1990, *The Practice of the Wild* (San Francisco: North Point Press), p. 178; Annie Dillard, 1989, *The Writing Life* (New York: Harper & Row), p. 68; Edwin Dobb, 1995, "Without Earth there is no heaven," *Harper's Magazine*, February, pp. 33–41; and Elizabeth Johnson, 1997, "Retrieval of the Cosmos in Theology," *Earthlight*, Spring, pp. 8–9.

66 The quotation by Heinrich D. Holland on "the relative dullness of Earth history" appears in his 1984 *The Chemical Evolution of the Atmosphere and Oceans* (Princeton: Princeton University Press).

67 Heinrich D. Holland and Ulrich Petersen, 1995, *Living Dangerously: The Earth, Its Resources, and the Environment* (Princeton: Princeton University Press).

68 John Maynard Smith and Eörs Szathmáry, 1995, *The Major Transitions in Evolution* (Oxford: Freeman).

69 For a full presentation of John Maynard Smith's views, see his 1984 "Science and Myth," *Natural History Magazine*, November, pp. 11–24.

70 Excerpts from many of Julian Huxley's books can be found in my 1994 *Evolution Extended: Biological Debates on the Meaning of Life* (Cambridge, Mass.: MIT Press). The paragraphs included here are drawn from p. 237, extracted from Huxley's 1957 *Religion Without Revelation* (New York: Harper & Row), pp. 26, 158.

Design elements at beginnings and ends of chapters are arrays of cottonwood buds, *Populus fremontii*.

Index

Abram, David, 122
Ackerman, Diane
  conversation with, 104–109
  mention of, 14, 159, 179, 264
Alien species
  as cause of extinctions, 97, 130, 262
  examples of, 112, 130–32
  humans as, 262
American Association for the Advancement of Science, 57
American Teilhard Association, 58–59, 65
Amythia, 49–51, 59, 71
Animal rights, 8, 254
Ants, 89, 254, 110
Appleyard, Bryan, 225–26, 263
Aprahamian, Maia, 78
Archaea, 172
Arthropods
  adaptation to terrestrial life, 41–42
  evolution of wings, 9, 42
  hive minds, 42
Atmosphere, gaian view of, 162–64, 169–74, 182–83, 190
Babbitt, Bruce, 261
Bacon, Francis, 16, 292
Bacteria, as foundation of nutrient cycles, 47, 186–91, 260
Balance-of-nature debate, 143–44
Barbour, Ian, 268
Barlow, Devon, 118
Barlow, Halsey, 117, 261–62, 271–72
Bateson, Gregory, 122, 204
Bateson, Mary Catherine, 20–30
Beaver, as keystone species, 127–29, 144
Berg, Peter, 122
Berry, Thomas
  biography of, 54
  mention of, 32, 52–55, 57–63, 73, 77–78, 217, 220, 226, 232, 237, 267–71
Binaries, metaphysical, 228–34
Biodiversity
  celebration of, 104, 119–20, 218
  credo, 241–42, 255–56
  crisis, 25–26, 93–94
  ecospiritual aspects, 105–106, 119–20, 217–21, 241–42

Biodiversity (cont.)
favorite organisms, 109–17, 211–12
fostering love of, 109, 119–20, *see also* biophilia
gaian aspects, 199–201, 258–61
hot spots, 255
increase through time, 47
number of species worldwide, 89
on islands, 97–99
recovery of Pleistocene mammals, 138–40
re-storying, 85–89
role in evolutionary epic, 85–87
ultimate value of, 151, 241–42
Biophilia
examples, 69, 108–20, 175, 211–12, 217
hypothesis, 7–9
Bioregionalism
credo, 242–43, 256–57
definitions, 122–23, 132
ecospiritual aspects, 145, 148–58, 218–21, 242–43
founding of, 16, 122–23
responsibility for endangered species, 123–24, 140–41
ultimate value of, 242–43
Biosphere
breathing of or with, 157, 224–25, 289, 293
closure in, 182–88
developmental history of, 190–200
lifespan of, 172, 174, 198
temperature history of, 192–200, 205
Bogs
as favorite habitat, 116–17, 205, 212
nitrogen fixers for, 259, 262
role in ice ages, 173, 205–206
Brachiopod, as living fossil, 86, 253
Bryophytes (mosses)
as favorite organism, 114–15, 149, 155, 219
role in bog formation, 117, 126
role in ice ages, 173, 206, 212
Buber, Martin, 264
Buddhism, 13, 16, 64, 67, 236, 264

Callicott, J. Baird, 123, 243, 251
Campbell, Joseph, 3, 61, 270, 292
Capra, Fritjof, 240
Carbon dioxide
cycling in atmosphere, 191, 224–25, 229
feedbacks in chemical weathering, 169–71
history of, 169–71, 181
Cathedral of St. John the Divine, 11–12, 119, 149–56
Christianity
bringing meaning out of suffering, 67, 70
fundamentalism in, 65, 276–77
green values in, 11–12, 25–26, 154–58, 236, 276
lack of green values in, 54, 58, 71, 152–55
ritual in, 74–75, 82, 154–58
Closed biosphere paradox, 184–90
Coelacanth fish, as living fossil, 86, 252
Coelho, Mary, 65–66
Coevolution, as alternative to Gaia theory, 167–68
Complexity, sciences of, 176, 240
Condor, reintroduction of, 137–39, 141
Confucianism, 60–61, 65
Conservation priorities, 140–41, 151, 248, 252–60
Copernicus, 31, 292

Cottonwood trees, 92, 127–30, 133, 138, 145, 219, 316
Council of All Beings, 13, 119–20
Credos
  examples of green, 46, 110, 236–37, 244–45, 246–47
  examples of traditional, 235–36
  for way of science, 236–61
Cyanobacteria, 153, 155, 196
Cycle v. arrow
  in evolution v. Gaia, 228–30
  symbol for science, 184–85, 190, 192, 199–200

Daisyworld model, 174–77, 204, 207–208, 214, 216
Darwin, Charles, 59, 160, 167, 182, 190, 238
Dawkins, Richard, 7, 38, 46, 66, 237, 287
Death, coming to terms with, 26, 67, 70–72, 138, 213–14, 231, 241, 265–66, 292
Deep ecology, 2, 214–16, 251, 266
Deep hot biosphere theory, 166
Dennett, Daniel, 28
Diamond, Jared, 8, 95–96
Dillard, Annie, 14, 237, 270
Disparity, taxonomic, 253
Dobb, Ed, 270–71
Dobzhansky, Theodosius, 37, 237, 279
de Duve, Christian, 46–47, 255
Dyson, Freeman, 11

Earth Day celebrations, 78, 118–19, 157–58
Earth star mushrooms, 80–81
Ecofeminism, 252, 265–66
Ecological restoration, 97, 138–40, 146–50, 257, 263
Ecology
  definitions, 121–22
  key ecospiritual principles, 149–50
Ecoreligious movement, 10–16
Ecosystem services, 140, 153–54
Eiseley, Loren, 30, 237
Eliade, Mircea, 18
Elk, extinct subspecies of, 123–24, 247
Endangered ecosystems, 125–26, 257
Endangered species
  bioregional responsibility for, 123–26, 256–57
  habitat fragmentation on, 99–101, 141
  hot spots, 125, 255
  management, 142–43, 255–56
  numbers of, 94
  our awareness of, 105
Endemic species, definition, 91
Environmental education, 1, 148–49
Environmental ethics, *see* morality
Epstein, Sam, 193
Equisetum (horsetails), 86, 229
Eukaryotic cell, origin of, 47, 196–97
Everglades, 129, 140–41
Everndon, Neil, 262
Evolution
  arms races, 191, 230–31
  convergent, 38–41
  cultural, 7, 53–54, 55, 64
  fact v. theory, 286–87
  initiative in, 231–32
  innovations in, 43, 66, 253–54
Evolutionary epic
  as basis for credos, 236–39
  for biodiversity recovery, 140–42, 145
  bioregional interpretation of, 122
  coining of term, 6, 24, 294
  contingency v. determinism in, 199–200, 232

Evolutionary epic (cont.)
continuance into future, 45–46, 241, 292
credo for pageant of life, 241–42, 252–55
criticism of rendering as mythic, 279–80
current versions, 30–57, 279
gaian views of, 211–21, 234
human role in, 30, 54–58, 76, 260–71
importance of, 23, 27–28, 49, 50–52, 64–65, 73, 120, 218–21, 293–96
IRAS conference on, 57, 79
meaning of, 225–26
meshing with religious traditions, 63–65, 75, 292–93
metaphors of, 30, 44, 241, 268–69, 296
moral values from, 26, 32–33, 50, 59, 69, 145, 236–42, 295
past versions, 28, 46, 59, 279
poetry for, 79
power of, 73, 227
prospects for success as worldview, 55–56, 293–96
as religion, 6, 23–28, 51–52, 56–57, 291–96
ritual for, 32, 77–83, 119–20, 292–93
six-way discussion of, 57–79, 240
trends and directionality interpreted in, 46–47, 190–200, 226–30
for way of science, 14, 54
as wisdom tradition, 51–52
Evolutionary psychology, 9
Exaptation, 9–10
Exotic species, *see* alien species
Extinctions, mass
causes, 177–81
end-Cretaceous, 91, 194
of plants, 90–91
Pleistocene, 95–96, 132–40, 248
pre-Pleistocene episodes, 47, 178
sixth major, 25, 90, 93–96, 100–101, 241–42, 259, 262

F_1 generation, definition of, 201, 209–10
Faint young sun paradox, 168–69
Faith
as basis for religion, 281, 287
for way of science, 287, 289–90
Feminism, 53, 68, 78, 252, 265
Foreman, Dave, 241, 255, 262
Fox, Matthew, 78
Fox, Warwick, 266
Francis of Assisi, 152–3

Gaia and Gaia theory
atmosphere role in, 162–64, 167–72, 180–81, 188
attraction for research, 202–209, 214–16
biodiversity importance for, 258–60
conference at Oxford University (1996), 195, 201, 210–11, 214, 273, 278
conference in San Diego (1988), 182
credo, 243, 258–60
criticisms of, 167–68, 173–76, 179, 187, 195–96, 205–206, 209, 272–73, 278, 288
cycle v. arrow as symbol of, 184–85, 190, 192, 199–200, 229–30
Daisyworld model, 174–77, 204, 207–208, 214, 216
definitions, 160, 198, 209, 218, 220
developmental view, 190–201
ecospiritual aspects, 14–15, 159–60, 175, 186, 198–201, 210–21, 228–34, 243–44, 267, 274, 285, 288–89, 293, 294

embedded self image, 267
examples of gaian mechanisms, 173, 205, 209
examples of research, 202–209
explanation for life's persistence, 166–82
homeostatic qualities, 166–67, 181, 200
human role in, 76, 263, 270–71
as hypothesis generator, 173, 210–11
landscapes crucial for, 258–60
metaphors of, 164, 186, 220
naming of, 162
natural selection in, 175, 187–88, 209
opposition to contingency viewpoint, 199–200, 232
origin of hypothesis, 159–66
paradoxes in, 168–69, 182–88
self-regulation in, 165, 172–76, 200, 218
symbiosis in, 186–87
Garrels, Bob, 207–208, 211
Geophysiological Society, 202, 273
Geophysiology, 159, 182
Ghost species, 20, 90, 132–37, 257
Gila River bioregion, 101–103, 123–33, 136, 142–44, 243–49, 257
Gila trout, 101, 123, 142–43
Gilligan, Carol, 252
Ginkgo tree, as living fossil, 87–88, 253
Glacial tillites, reinterpretation of, 194–95, 197
Global warming, 259
Gold, Thomas, 166
Golliher, Jeff, 150
Good and evil, 28
Goodenough, Ursula
biography of, 57, 62–63
conversation with, 62–78, 286–88
mention of, 44, 223, 237, 241, 289,
Goodwin, Brian, 217
Gould, Stephen Jay, 11, 17, 44, 225–26, 232, 238
Grassie, Billy, 79–80
Grazing of livestock, 143, 247
Green space, green time, 15, 218–19, 221, 234
Grim, John
biography of, 57–60
conversation with, 60–73
Ground sloths, 94, 96, 115, 134–36, 226, 256
Gustafson, James, 3

Habitat fragmentation, 99–102, 141–42, 147, 255
Haldane, J. B. S., 203
Hanh, Thich Nhat, 13
Harding, Stephan
biography of, 214–16
conversation with, 110–12, 214–18
mention of, 118, 223, 266, 270
Hardy, Alister, 38, 237
Harjo, Joy, 79, 227
Harvey, William, 182
Haught, John, 237
Hefner, Phil, 80
Hinkle, Gregory, 186
Hoffert, Martin, 198
Holland, Dick
conversation with, 272–77
mention of, 281–82
Horseshoe crab, as living fossil, 86, 252–53
Houston, Jean, 292
Hudson River bioregion, 124, 146–51, 158
Humanities
role in evolutionary epic, 26–28, 76–77

Humanities (cont.)
two cultures conflict, 35, 53, 77, 279–81
Humans, role of
alien species, 262
biotransporters, 189
cause of sixth great extinction, 94–96, 105, 137, 145, 241, 248, 262
celebrants of the Universe Story, 53–57, 76, 270–71
as chimpanzee, 52
construction of role, 227, 269
keystone species, 130
life/evolution/universe become conscious, 30, 56, 269–71
meaning-makers, 56, 250, 261–72
meteor impact, preventers of, 263
stewards of Earth, debate over, 150, 216–17, 236, 275
Vernadsky's view of, 189–90
Huxley, Aldous, 267
Huxley, Julian
biography of, 294
conversation with (imagined), 281–83, 293–96
mention of, 28, 30, 46, 53, 57, 59, 107, 220, 237, 270, 283–84

Ice ages
disputing past occurrences of, 194–95, 198
gaian causes of, 206, 209
Indigenous cultures
as cause of extinctions, 94–96, 137
knowledge of nature, 89
religions, 13, 53, 59–60, 64, 74, 117–18
Institute on Religion in an Age of Science, 57, 63, 80
Island biogeography, theory of, 24, 98–99
Islands, vulnerability to extinctions, 97
Ivory-billed woodpecker, 90

James, William, 15, 235
Jaw, origin of, 9–10
Jeffers, Robinson, 79, 122, 236
Johnson, Elizabeth, 270–71
Joint Appeal by Religion and Science for the Environment, 11
Jung, Carl, 61

Kant, Immanuel, 2
Kauffman, Stuart, 17, 176
Kaufman, Gordon, 30
Kelly, Kevin, 44
Kestrels, 89, 114
Keystone species, 126–30, 256
Kimura, Motoo, 11
Klinger, Lee
biography of, 202, 205–206
conversation with, 116–17, 205–13
mention of, 219–20
Knauth, Paul, 193
Koenigsberg, Betsy, 102, 118
Koenigsberg, Steven, 118
Koestler, Arthur, 267
Kohlberg, Lawrence, 3
Komodo dragon, 97–98, 254
Kump, Lee
biography of, 202, 207–209
conversation with, 207–12

Lapo, Andrei, 237
Lenton, Tim
biography of, 201–203
conversation with, 202–13
Leopold, Aldo, 243

Lichens
convergent evolution of, 41
as favorite organism, 211
as living fossil, 87, 89
role in chemical weathering, 171, 204
Liebig's Law of the Minimum, 150
Life
distinction from nonlife, 223–24
gaian view of, 162–63, 224–25
persistence of, 166–82
Living fossils, 86–87, 252–53
Logging, 116–17
Lopez, Barry, 14
Lovelock, James
conversation with, 161, 175
mention of, 29, 160–68, 173–75, 186, 190, 195, 200–17, 237, 244, 270, 272, 275, 278
Lovelock, Sandy, 161
Lysenko, T. D., 278–79

MacArthur, Robert, 24, 98
Macy, Joanna, 13, 119, 265
Mammals
marsupial v. placental, 38, 87
monotremes, 87
total number of, 89
Mandalas for science, 184–85, 233–34
Mankiewicz, Paul and Julie
conversation with, 113–15, 146–56
mention of, 275
Margulis, Lynn
conversation with, 112–13
mention of, 11, 186–90, 212, 215, 237, 258–59, 272
Mars, looking for life on, 161–62, 165–66
Martin, Paul, conversation with, 115, 133–40
Matthews, Cliff, 233
Maynard Smith, John, 46, 195–96, 237–38, 277–81, 285, 293
McKay, David, 165
McMenamin, Mark, 196, 237
Meaning
construction of, 4–7, 15, 24–30, 56–57, 66–67, 215, 223–28, 240, 271, 279–80
credo, 250, 261
emergence of in life, 44–45, 56, 223–26, 261
enrichment of, 225, 250, 261–72
gaian construction of, 188, 210–20
v. purpose, 226
Memes, 7, 55–56
Mesquite tree, 42, 132, 140
Metaphor
for evolutionary epic, 30, 44, 268–69, 296
importance of, 34
root metaphors, 268
Meteor impacts, 177–79, 194–95, 263–64
Microbial mats, role in Gaia, 187–88, 212
Migrations as endangered phenomena, 101–105, 141–42, 254
Mills, Stephanie, 122
Mitchell, Edgar, 159
Mitchell, Joni, 33
Monarch butterflies, 20, 101–105, 147, 219
Monod, Jacques, 3–6
Moore, John, 62
Morality, ecological
alien species, 112, 130–32
cosmological grounding of, 50, 58–59, 237, 295
credos, 236–44
credos applied, 244–61

Morality, ecological (cont.)
examples of, 108, 110, 114–17, 215–16
for biodiversity, 26, 98, 138–40, 241–42, 255–56
for bioregionalism, 122–24, 144, 151–52, 242–43, 256–57, 258–60
for pageant of life, 32, 62–63, 69, 71, 76, 145, 240–41, 252–55
for Gaia, 243, 258–60
for wilderness preservation, 108, 145, 263
fostering, 6–16
human culpability for extinctions, 96, 105, 137, 145, 241, 248, 262
importance of cultivating, 1–3, 20–21, 50
misanthropism, countering, 250, 262–64, 269, 271
origin of, 50, 58–59, 276, 279
religious conviction for, 20–21, 50, 244
self-image effects on, 264–71
Morality, human to human, 238
Morton, James Parks, 12, 155–56
Muir, John, 14
Multiculturalism, 50, 63–65
Muntjac deer, 111, 216, 223
Mussels, 125, 129–31
Myers, Norman, 25, 255
Myth, as grand narrative, 23–24, 50, 237, 279–80
Mythopoeic drive, 5, 7–11, 18, 24, 48

Naess, Arne, 2, 214, 235, 266, 270
Narratives, role of, 23–24, 48, 69, 237, 279–80
National Religious Partnership for the Environment, 12
Native American religions, 13, 16, 60, 74, 117–18
Natural selection
as cause of religion, 4
in Gaia theory, 175, 187–88, 209
Nelson, Richard, 13–14
Nietzsche, Friedrich, 2, 156
Nitrogen, cycling of, 184–85, 200, 203, 229
Noss, Reed, 125, 258

Overkill hypothesis, 95–96, 115, 137
Overpopulation, 260–61, 269
Oxygen
origin in atmosphere, 180, 199, 289
role in wildfires, 207, 209

Paganism, 12–13, 75, 80, 82, 277
Parasites, role of, 92, 129–30
Passenger pigeon, 92–93
Perception
in humans for evolutionary epic, 250, 269–71
in humans for Gaia, 270
in life, 89–90, 225–26, 253–54
as prior to science, 217
Peregrine falcon, reintroduction of, 124, 147–48
Personal wholeness, 69, 72–73, 281
Photosynthesis, ecospiritual rendering of, 32
Plants
drought resistance, 40, 42–43
mass extinctions, 88–89
mobility of seeds, 43
reminders of ghost herbivores, 132–33, 140
Plastids, convergent evolution of, 40–41
Pluralism, importance of, 15, 64, 240, 260
Pollinators, importance of, 91
Postmodernism, 290

Predation, 49, 231
Protoctists, 112–13, 187
Purposefulness
cosmic lack of, 33–34, 56
gaian lack of, 174–75
limited to life, 44–45, 223–26

Rader, John, 235
Rampino, Michael, 177, 194–96, 198
Raven, Peter, 11, 92
Rawls, John, 257
Redfield, Alfred, 203
Religion
criticism of, 2–3
definitions, 3
evolutionary epic as, 6, 23–28, 51–52, 56–57, 291–96
origin of, 4–7, 10
resistance to change, 282–85
separatism from science debate, 272–96
supernatural aspects, 3, 51
Religious naturalism
definition, 3
examples, 106–108, 175, 210–20, 282–96
Resourcism, 238
Restoration ecology, 114, 127
Ritual
for evolutionary epic, 32, 75–76, 78–83, 157–58, 293
for Gaia, 289, 293
green, 12, 152–58
importance of, 74–76, 292–93
see also Earth Day *and* winter solstice
Rolston, Holmes, 224, 242, 265, 270
Rue, Loyal
biography of, 57–58
conversation with, 51–52, 58–78, 83, 283–86
mention of, 3, 49–51, 233, 237, 240, 265, 296

Sacred, sense of, 9, 11–12, 16, 54, 107, 151–52, 244, 282
Sagan, Carl, 11, 159
Sagan, Dorion, 237
Salthe, Stanley, 176
Schaible, Bob, 79
Schneider, Stephen, 11
Schumacher College, 215–17
Schwartzman, David
biography of, 201, 203–204
conversation with, 192–200, 203–13
mention of, 171–72, 177, 229
Science
community effort, 18
mandalas for, 184, 233–34
openness to change, 17–19, 282–85, 289–93
openness to interpretation, 17
postmodern criticism of, 290
practice of, 18
Yoga of the West, 55
see also way of science
Scientific materialism, 6
Scientism, 53
Secular humanism, 2
Seed, John, 13, 119–20
Self
communal image of, 267
construction of image, 269
embedded image of, 267
expanded image of, 2, 120, 215–16, 257–58, 265–67, 270
illusion of, 67
isolated image of, 264
relational image of, 264–65
Selfish gene theory, 66
Separatism of science and values, 238

Sex, origin of, 92, 265–66
Sexual selection, 10, 232
Shepard, Paul, 122
Simpson, George Gaylord, 39
Smith, Huston, 3
Snyder, Gary, 16–17, 121–22, 159, 201, 241–42, 270
Social coherence, 4, 26, 69, 73, 76
Sociobiology, 4–5, 9, 24
Spencer, Herbert, 238
Spinoza, 266
Starfish, as keystone species, 129
Stewardship of Earth, 12, 150, 216–17, 236, 275–76
Storytelling
  importance of, 23–28, 48, 120, 291–92
  to evoke the evolutionary epic, 77, 119–20
Sturges, Philemon, 36, 82
Sun
  death of, 72
  role in evolutionary epic, 32
Swimme, Brian
  biography of, 57–58, 61–62
  conversation with, 35–36, 52, 61–79
  his version of evolutionary epic, 31–36, 49
  mention of, 50–51, 55, 83, 219, 225, 237, 268, 270–71, 295
Symbiogenesis, 41, 46, 231
Symbiosis, 48, 89–90, 186–87, 230–32, 254, 259
Synergy, 230–31
Szathmáry, Eörs, 46

Tanager, 20, 100, 109, 118, 123
Taoism, 13, 65, 233
Teilhard de Chardin, Pierre, 28, 54, 57, 58, 60–61, 72, 191, 220, 279
Teleology, *see* purposefulness
Templeton Foundation, 286
Thermophiles, 172, 195–98
Thoreau, Henry David, 14
Tiamat
  character in Sumerian myth, 292
  name for supernova, 33–35, 56, 82, 225
  ritual, 82–83, 292–93
Tickell, Crispin, 294
Transpersonal ecology, 252
Trophic cascade, 138
Tucker, Mary Evelyn
  biography of, 57–61
  conversation with, 60–77, 83
  mention of, 219, 296

Umbrella species, 255–56
Unitarian-Universalism, 74–75, 118–19, 157, 277

Value
  anthropocentric v. ecocentric, 2, 238–39, 244–50
  ecofeminist view of, 252
  expanding the circle of, 258
  genesis of, 237–38, 252, 280
  intrinsic, 215, 251
  pluralism in, 240, 260
  separation from science, 238, 276–81
  ultimate, 151, 235–38, 251
  veil of ignorance, 257
  see also morality
Vernadsky, Vladimir, 188–92, 200, 203–204, 229, 237
Volk, Dana and Eliza, 82
Volk, Tyler
  biography of, 202, 204–205
  conversation with, 204–12, 288–93
  mention of, 171–72, 177, 182, 192, 196, 267

Wallace, Alfred Russel, 182
Watson, Andrew
biography of, 202, 206–207
conversation with, 206–14
mention of, 175
Way of ancients, 12–13
Way of immersion, 13–14
Way of reform, 11–12, 75, 292–93
Way of science
credos, 236–44
criticisms of, 272–81
for contemplating metaphysical binaries, 140–43
for greening worldviews, 14–15, 54, 96–97, 233–34
openness to change, 17, 19, 107, 282–85, 288–92
overall crafting of, 14–21, 218–21
religious language for, 107–108
story of the changing story, 291–93
symbol for, 234
underpinned by evolutionary epic, 85, 237
Way of transcendence, 13
Weathering, chemical, 169–71, 193, 196, 205
Weinberg, Steven, 66, 225
Weisskopf, Victor, 11
Whitehead, Alfred North, 61
Whitman, Walt, 14, 43, 79, 83
Wicken, Jeffrey, 237
Wilderness
as arena of evolution, 145, 239, 247, 250, 255
biological impoverishment of, 133, 137–40, 147
fragmentation of, 259
grazing on, 249
management of, 142–45, 246–47
origin of protected status, 243
recovery of, 145, 263
religious feelings for, 108, 145
Wildlands Project, 145
Williams, George C., 196
Williams, Terry Tempest, 146
Wilson, Edward O.
biography of, 24–25
conversation with, 23–27, 48, 56–57, 110
mention of, 4–9, 30, 46–47, 53, 59, 98, 109, 199, 237, 242, 269, 294
Windsor, Donald, 92
Wings
convergent evolution of, 39, 42
origin for insects, 9, 42
Winter, Paul, 119
Winter solstice celebrations, 75, 82, 119
Wisdom traditions, 51–52
Wolf, reintroduction of, 124, 138–39, 244–60
Wordsworth, William, 14
Wuerthner, George, 143